Innovations in Ground Water and Soil Cleanup

From Concept to Commercialization

Committee on Innovative Remediation Technologies
Water Science and Technology Board
Board on Radioactive Waste Management
Commission on Geosciences, Environment, and Resources
National Research Council

NATIONAL ACADEMY PRESS
Washington, D.C.   1997

NATIONAL ACADEMY PRESS • 2101 Constitution Avenue, N.W. • Washington, DC 20418

NOTICE: The project that is the subject of this report was approved by the Governing Board of the National Research Council, whose members are drawn from the councils of the National Academy of Sciences, the National Academy of Engineering, and the Institute of Medicine. The members of the committee responsible for the report were chosen for their special competences and with regard for appropriate balance.

This report has been reviewed by a group other than the authors according to procedures approved by a Report Review Committee consisting of members of the National Academy of Sciences, the National Academy of Engineering, and the Institute of Medicine.

Support for this project was provided by the U.S. Environmental Protection Agency under Agreement No. CR 82307, the U.S. Department of Energy under Agreement No. DE-FC01-94EW54069, and the Department of Defense under DACA87-94-C-0043. Any opinions, findings, conclusions, or recommendations expressed in this publication are those of the author(s) and do not necessarily reflect the view of the organizations or agencies that provided support for the project.

Library of Congress Cataloging-in-Publication Data

Innovations in ground water and soil cleanup : from concept to
      commercialization / Committee on Innovative Remediation
      Technologies, Water Science and Technology Board, Board on
      Radioactive Waste Management, Commission on Geosciences,
      Environment, and Resources, National Research Council.
            p. cm.
      Includes bibliographical references (p.   ) and index.
      ISBN 0-309-06358-2
      1. Hazardous waste site remediation. 2. Hazardous waste
      management industry—Technological innovations. I. National
      Research Council (U.S.). Committee on Innovative Remediation
      Technologies.
      TD1030.I56 1997
      628.5—dc21   97-21190

Innovations in Ground Water and Soil Cleanup: From Concept to Commercialization is available from the National Academy Press, 2101 Constitution Avenue, NW, Lockbox 285, Washington, DC 20055 (1-800-624-6242; http://www.nap.edu).

Cover art by Y. David Chung. Chung is a graduate of the Corcoran School of Art in Washington, D.C. He has exhibited his work throughout the country, including at the Whitney Museum in New York, the Washington Project for the Arts in Washington, D.C., and the Williams College Museum of Art in Williamstown, Massachusetts.

The National Academy of Sciences is a private, nonprofit, self-perpetuating society of distinguished scholars engaged in scientific and engineering research, dedicated to the furtherance of science and technology and to their use for the general welfare. Upon the authority of the charter granted to it by the Congress in 1863, the Academy has a mandate that requires it to advise the federal government on scientific and technical matters. Dr. Bruce M. Alberts is president of the National Academy of Sciences.

The National Academy of Engineering was established in 1964, under the charter of the National Academy of Sciences, as a parallel organization of outstanding engineers. It is autonomous in its administration and in the selection of its members, sharing with the National Academy of Sciences the responsibility for advising the federal government. The National Academy of Engineering also sponsors engineering programs aimed at meeting national needs, encourages education and research, and recognizes the superior achievements of engineers. Dr. William A. Wulf is president of the National Academy of Engineering.

The Institute of Medicine was established in 1970 by the National Academy of Sciences to secure the services of eminent members of appropriate professions in the examination of policy matters pertaining to the health of the public. The Institute acts under the responsibility given to the National Academy of Sciences by its congressional charter to be an adviser to the federal government and, upon its own initiative, to identify issues of medical care, research, and education. Dr. Kenneth I. Shine is president of the Institute of Medicine.

The National Research Council was organized by the National Academy of Sciences in 1916 to associate the broad community of science and technology with the Academy's purposes of furthering knowledge and advising the federal government. Functioning in accordance with general policies determined by the Academy, the Council has become the principal operating agency of both the National Academy of Sciences and the National Academy of Engineering in providing services to the government, the public, and the scientific and engineering communities. The Council is administered jointly by both Academies and the Institute of Medicine. Dr. Bruce M. Alberts and Dr. William A. Wulf are chairman and vice-chairman, respectively, of the National Research Council.

Preface

Analysts have estimated that the total cost of cleaning up some 300,000 to 400,000 contaminated sites, located on both public- and private-sector facilities, could reach approximately $500 billion to $1 trillion (see Chapter 1). Even if such a staggering cost were indeed incurred, there is no guarantee that available technologies will clean up all of these sites to meet legal requirements. Furthermore, innovative remediation technologies that hold considerable potential for providing enhanced soil and ground water cleanup are infrequently selected by waste site managers, remediation consultants, site owners, and regulators—all of whom may be risk averse and concerned about the possible failure of new technologies to deliver on their potential. In many cases, government agencies and corporations spend large sums on new technology development and testing, but site managers do not select the new technologies at federal or corporate sites. This paradoxical situation has produced a considerable debate among all parties concerning how to fix the waste site remediation problem.

Two types of broad solutions to the problems of contaminated site remediation are receiving increasing attention. First, there are increasing attempts to prioritize sites that need immediate attention and then to reconsider the remediation end points based on site-specific risk assessments at sites judged to have low risks. A relaxation of cleanup goals is being sought at some sites. Some observers perceive this strategy as remediation of the regulations rather than the contaminated sites. Nevertheless, this risk-based, site-specific approach is increasingly popular among both government agencies and private companies confronted with budgetary constraints. But, several questions about the validity of this approach and how to implement it remain unanswered: Who is engaged in the process of determining the criteria by which the sites are prioritized? How do we determine these less-stringent end points? How do we differentiate between risks as determined by professional risk assessors or by site owners and those perceived by residents near contaminated sites?

The second type of solution being explored, and the one addressed in this report, is to promote the development and increased use of innovative remediation technologies. Attempts at remediation have historically favored established technologies: pump-and-treat systems for contaminated ground water and excavation followed by incineration or disposal for contaminated soil. Some would say that even though waste cleanup regulations per se do not specify the type of technology that must be used to meet regulatory cleanup requirements, implementation of the regulations has resulted in a "technology push" paradigm: that is, established technologies are used because they have been used before, so their performance characteristics are well known. Experience over the past two decades has revealed a repeating pattern at many sites, suggesting that this approach has led to less-than-optimal cleanups. The challenge is to create a new policy strategy that marshals appropriate economic and regulatory drivers to encourage innovation in ground water and soil cleanup and better environmental stewardship.

This new approach for site remediation is based on a shift to a policy paradigm that relies on market demand rather than technology push. That is, the market (i.e., client needs for site cleanup) generates the strong forces necessary to propel remediation technology development and commercialization. The current market tends to force technology developers and service providers to seek out reluctant, risk-averse customers and investors. Instead, in the new market, clients (i.e., all types of remediation technology users) actively seek solutions based on new remediation technologies. The primary goal of remediation technology development under this new paradigm is to continually increase the diversity and number of technologies included on the menu of options considered by site owners, regulators, and consultants. Testing at several sites using consistent protocols and making cost and performance data available for peer review comprise the essential elements of technology development. Also, under the new paradigm the various stakeholders, particularly the concerned public living near contaminated sites, must be engaged very early in the evaluation of technologies for site cleanup.

This report summarizes the extended deliberations of a committee of experts in contaminated site remediation and innovative technology commercialization. The 16 committee members represented a balance of viewpoints and included a representative of a public interest organization active in site remediation, a patent attorney, a venture capitalist, and a technology developer, as well as technical experts from universities, environmental consulting firms, and industry. At the committee's six meetings, invited guests representing government agencies, site owners, and technology developers presented to the committee their perspectives on constraints and opportunities for using innovative technologies for site remediation. To these colleagues, who took time from their busy schedules to speak to us and provide valuable follow-up materials, the committee is extremely grateful.

The diversity of backgrounds and expertise of the committee members and the wide range of opinions held by these members meant that my job as committee chair was to ensure that in its deliberations, the committee moved forward toward its final goal: a consensus view of what might be the new policy paradigm for selection of remediation technologies at sites with contaminated ground water and soil. During our debates, some committee members served as strong forces that pulled the committee in new directions. Other members provided the moderating influences. Still others ensured that the allure of new remediation technologies and financial incentives did not obscure our ultimate goal: responsible environmental stewardship.

When asked to chair the committee, my immediate concern was that my research and teaching experience, which is focused on remediation technology development and testing in an academic setting with only an occasional foray into the consulting world, provided only half the expertise needed to guide our deliberations on commercialization of remediation technologies. It was clear that the committee required the experience and wisdom of a colleague with considerable experience in the "real world" to provide the other half of the committee's leadership. Dick Brown agreed to co-chair the committee to offer his practical experience gained from two decades of developing and implementing remediation technologies at a large number of contaminated sites. I appreciate his advice and support.

Another element responsible for the success of this project was the open-mindedness of committee members. Committee members not only articulated their own ideas and positions forcefully, but they also were able to listen objectively to others' ideas and positions and, based on these, were willing to transform their arguments into a consensus. I was fortunate to work with a committee in which the members were willing to listen to and thoughtfully consider others' opinions while maintaining a sense of humor if their own suggestions were not readily adopted.

Given the many distractions of the committee members' daytime, paid jobs, it was not always easy to deliver on the commitments made during the inspiring moments of committee meetings. Thus, a disciplined organizer was essential for the successful conclusion of the committee's activities. Jackie MacDonald, the Water Science and Technology Board (WSTB) staff officer who worked with the committee, ensured that the many exciting discussions at meetings were translated into written documents that could be reviewed by others and edited to produce a coherent document. But, Jackie was much more than a passive, behind-the-scenes coordinator; she actively participated in all of our discussions, and she offered insightful comments and input. She repeatedly edited our written contributions to crystallize a logical flow of ideas and to maintain consistency in our arguments. That she carried out this arduous task while under the considerable stress of a family health crisis is a testimonial to her professionalism and dedication. Angie Brubaker provided the essential administrative support necessary for organizing committee meetings. She also used her considerable organizational and production skills to coordinate the preparation of the many drafts of this report. All of the committee members appreciated, as I do, Angie's help throughout the two-year study period.

Several sponsors were early believers in this project and elected to provide the generous financial support needed to launch this study. These sponsors discerned the need for a follow-on study to the 1994 WSTB report *Alternatives for Ground Water Cleanup* (chaired by Mike Kavanaugh and staffed by Jackie MacDonald). On behalf of the WSTB and the committee, I thank Richard Scalf (retired) and Stephen Schmelling of the Environmental Protection Agency's R. S. Kerr Environmental Research Laboratory; Sherri Wasserman Goodman, deputy under secretary for environmental security at the Department of Defense, and Colonel James Owendoff, formerly with the Department of Defense; and Clyde Frank, deputy assistant secretary for technology development at the Department of Energy and Gary Voelker, Stephen C. T. Lien, and Stanley Wolf of the Department of Energy for their early insights and strong support of this project. *Alternatives for Ground Water Cleanup* was widely popular among government agencies, private-sector companies, remediation practitioners, and academic researchers. We can only hope that the standards set by that earlier WSTB report can be met by our efforts.

As difficult and time-consuming as Naional Research Council committee activities can be, the reward at the end is always worth the effort: a confluence of diverse ideas of acknowledged experts, so that consensus advice is provided on how good science can influence regulations and serve public policy needs in a timely fashion. I am glad that I had yet another opportunity to participate in this rewarding process.

P. S. C. Rao
University of Florida
Gainesville, Florida

Postscript: The following poem, which I composed for an early committee meeting, provides a vision of the goals sought by those working to develop and commercialize new remediation technologies.

> Imagine for a moment,
> a perfect world.
>
> A perfect world of remediation in which
> there was no need for regulatory push,
> the PRPs always take the high ground
> to clean up sites voluntarily,
> and they do not litigate to delay;
>
> A perfect world of remediation in which
> the sources can always be found with certainty,
> and the contaminant plumes
> always self remediate intrinsically
> or the presumptive remedy was
> indeed the best technology for the site;
>
> A perfect world of remediation in which
> the stakeholders' concerns were
> always addressed early and often,
> there were economies of scale up,
> there was no Valley of Death,
> and the investors always made enough profits.
>
> Now, wake up,
> stop imagining
> and look around carefully
> remembering that imagining
> an ideal world is just an escape,
> from the real world.
>
> But, before you despair,
> ask yourself how and act to
> transform the real world
> into the one you just imagined.

Contents

Executive Summary

U.S. taxpayers and corporations are spending large sums of money for the cleanup of contaminated ground water and soil at hundreds of thousands of waste sites across the country. Business analysts estimate that spending on waste site cleanup totaled $9 billion in 1996 alone. Federal accountants estimate that taxpayers will spend between $234 and $389 billion over the next 75 years for the cleanup of contaminated sites on land owned by the Departments of Defense, Energy, Interior, and Agriculture and the National Aeronautics and Space Administration. Estimates have placed the total cost of cleaning up all contaminated sites, privately owned and publicly owned, as high as $500 billion to $1 trillion.

Despite the large sums invested, the problems associated with waste site cleanup are far from solved. Conventional technologies, especially those for cleaning up contaminated ground water, have been unable to restore many types of sites to the standards set by environmental regulations for protection of public health and the environment. For example, the 1994 National Research Council (NRC) report *Alternatives for Ground Water Cleanup* evaluated the performance of conventional pump-and-treat systems for ground water cleanup at 77 sites and found that regulatory standards had been achieved at only about 10 percent of the sites. The limitations of conventional ground water cleanup technologies are now widely recognized by environmental engineers, scientists, regulators, and others involved in waste site remediation.

The inadequacy of conventional remediation technologies, along with the high costs of remediation, is in part responsible for increasing pressure on policymakers to limit the number of contaminated sites that are actively cleaned up. Yet, leaving contaminants in place rather than cleaning up sites involves costs and uncertainties that are not always recognized. Those responsible for the contaminated site must bear the cost of continued liability should contamination escape from the site into surrounding communities. Predicting the potential for such contaminant migration off site is subject to significant uncertainties, so that the full costs of this long-term liability are difficult to calculate. Further, maintaining the site to prevent exposure to the contamination may involve long-term costs. Costs are also associated with decreased property values and difficulty in selling property when significant contamination remains in place. Finally, leaving contamination in place is unacceptable to members of some communities near contaminated sites. Affordable remediation technologies that can remove the bulk of contaminant mass from the subsurface at contaminated sites are needed to reduce the long-term risks, liabilities, and costs associated with these sites.

While technologies for waste site cleanup have advanced in recent years, the menu of cost-effective options and use of existing innovative technologies are still limited, especially for contaminated ground water at large, complex sites. Innovative technologies have been selected for ground water restoration at just 6 percent of sites regulated under the Superfund program for contaminated site cleanup, according to recent Environmental Protection Agency (EPA) data. A noncomprehensive review of contaminated sites regulated under the Resource Conservation and Recovery Act (RCRA) found that innovative technologies were being used for ground water cleanup at 13 percent of the sites. A recent General Accounting Office audit of contaminated federal facilities found that "few new technologies have found their way into cleanups."

This report analyzes options for stimulating development and commercialization of technologies for reducing the costs and improving the effectiveness of ground water and soil cleanup at contaminated sites. It focuses on technologies that treat contaminated ground water in place in the subsurface and technologies that treat contaminated soil directly at the site, either in place or in a treatment unit. The report suggests ways to strengthen market forces to create demand for innovations in these types of technologies; reviews the status of remediation technology development, identifying where technology needs are greatest; outlines criteria that should be used to assess remediation technology performance; describes strategies for testing remediation technologies to measure their performance against these criteria; and recommends methods for comparing the costs of alternative remediation technologies.

The report was written by the NRC's Committee on Innovative Remediation Technologies, appointed in 1994 to develop testing and performance standards for subsurface cleanup technologies and to examine other issues related to commercialization of these technologies. The committee consisted of experts in hydrogeology, soil science, environmental engineering, environmental policy, patent law, finance, and public opinion. The committee's findings, as reported in this study, are based on reviews of technical literature and government reports; consultations with a range of stakeholders involved in waste site remediation, including federal and state regulators, industry groups, heads of start-up remediation technology companies, and venture capitalists; and the expertise of the committee members.

STIMULATING THE MARKET FOR
INNOVATIVE REMEDIATION TECHNOLOGIES

Despite the billions of dollars being spent on environmental cleanup each year, companies founded on marketing new environmental remediation technologies have fared poorly in almost all cases. Of seven companies that have gone public based on marketing a technology for waste site cleanup, the stock value of six of the companies has dropped since the initial public offering. Although there has been a healthy level of scientific research aimed at developing new environmental remediation technologies, with patent applications in this area increasing from nearly zero in 1980 to more than 430 in 1993–1994, few of these new scientific ideas have been successfully commercialized. Venture capitalists who could provide the critical funding for moving discoveries from the laboratory to commercial application have generally shied away from the waste site cleanup industry. For example, while total venture capital disbursements to all industries have more than doubled since 1991, venture capital disbursements to environmental technology companies have decreased by more than half during this period.

A major failing of national policy in creating a healthy market for environmental remediation technologies is the lack of sufficient mechanisms linking the prompt cleanup of contaminated sites with the financial self interest of the organization responsible for the contamination. Several large corporations evaluated in this study spend an average of about 5 percent of their earnings on waste site remediation, yet corporate managers rarely look to innovative remediation technologies as a means of reducing costs. Under the current system, especially at sites regulated under the Superfund and RCRA programs, it is frequently perceived as more cost-effective for responsible parties to delay cleanup than to install an innovative cleanup system. Even when regulations require site cleanup, the implementation process is long and is easily extended by review and appeals. For example, from a financial perspective, incurring annual costs of $1 million for litigation to delay cleanup at a contaminated site is more cost effective for many companies than initiating a cleanup that might require a $25 million cash outlay. Enforcement of waste site cleanup regulations is inconsistent, so the risk of a major penalty for delaying cleanup is low. Adding to the incentive to delay is the possibility that legislative reforms will relax the requirements for site cleanup at some future date, making it financially unwise to invest in technologies to reach today's more stringent cleanup goals. Companies that make an effort to initiate waste site cleanups promptly may be placed at a competitive disadvantage when compared to competing companies that delay cleanup. The result of the failure to link prompt cleanup of contaminated sites to corporate financial self interest is low demand in the private-sector market for environmental remediation technologies that can achieve improved performance at lower costs. Environmental remediation technologies are more of a legal product than a technological one, and there is little or no premium for improved solutions to subsurface contamination problems.

In the public-sector environmental remediation market, inadequate cost containment has contributed to delays in remediation and has decreased incentives for selecting innovative remediation technologies. According to the General Accounting Office, federal remediation contractors are often placed on "auto pilot" after being awarded a cleanup contract on a cost-reimbursable basis. With no incentive to contain costs, and in fact an incentive not to do so, quick action and cost effectiveness in remediation technology selection go unrewarded.

Delays in waste site remediation occur not just because of lack of financial incentives for prompt action but also because of the long time period that can be required to obtain regulatory approval of a cleanup plan and because of technical uncertainties. According to a Congressional Budget Office review, the average time between the proposal for listing a site on the Superfund National Priorities List and construction of the cleanup remedy is 12 years. While delays due to technical uncertainties are unavoidable, delays due to slow action by site owners and slow approvals by regulators need to be controlled in order to revitalize the market for innovative remediation technologies. Start-up remediation technology development companies have gone out of business while awaiting all of the final approvals necessary to use their technology at a large enough number of sites to stay solvent. Delays in environmental remediation discourage investment in these start-up companies due to the inability to predict the timing of investment returns.

Other factors also contribute to the weakness of the remediation technologies market and the poor success record of new remediation technology ventures. These include unpredictable time lines for remediation technology selection, which prevents technology developers and investors from accurately projecting investment returns; lack of consistent regulatory standards among various regulatory programs (Superfund, RCRA, underground storage tank, and state) and even within programs, making it difficult for technology developers to assure potential customers that their technology will meet regulatory approval; and lack of market data due to reluctance to disclose information about the magnitude and nature of site contamination problems, precluding the development of accurate market assessments for new technologies.

Recommendations: Stimulating Markets

To stimulate the market for innovative ground water and soil cleanup technologies, the committee recommends a variety of initiatives. Some of these are targeted at creating strong links between the financial self interest of those responsible for site contamination and the rapid initiation of site cleanup activities.

Others are intended to increase the certainty of the regulatory process so that technology developers and investors can more accurately predict their potential investment returns. (See Chapter 2 for a detailed description of these strategies.)

• **The U.S. Securities and Exchange Commission (SEC) should clarify and strictly enforce requirements for disclosure of environmental remediation liabilities by all publicly traded U.S. corporations.** Clarifying the existing requirements for reporting of environmental liabilities and strictly enforcing these requirements would provide an incentive for companies to initiate remediation, rather than delaying it, in order to clear their balance sheets of this liability. Detailed accounting procedures for complying with this requirement, along with a mechanism for certifying environmental accountants, need to be established by the U.S. accounting profession, possibly using the model of the International Standards Organization's series of standards for environmental management systems. Although technical uncertainties will preclude exact computations of remediation liabilities, companies should nonetheless be required to report their best estimates of these liabilities using reasonable estimates of probable remediation scenarios.

• **The SEC should enforce environmental liability reporting requirements through a program of third-party environmental auditing.** The possibility of an environmental audit, along with strong penalties for failing the audit, would help ensure that companies would comply with SEC requirements to report environmental liabilities. Certified public accountants, ground water professionals, or all of these could conduct the audits after receiving appropriate training.

• **Congress should establish a program that would allow companies to amortize the remediation liabilities they report over a 20- to 50-year period.** Such a program would ensure that by fully evaluating and disclosing their remediation liabilities with the best available current information, companies would not risk losing a major portion of their asset value. It would also provide a measurable cost target for remediation technologies to beat (the total cost of the declared liabilities).

• **The EPA should work to improve enforcement of Superfund and RCRA requirements.** Consistent, even-handed enforcement is essential for ensuring that U.S. companies are not placed at a competitive disadvantage compared to their domestic competitors by spending money on environmental remediation.

• **Managers of federally owned contaminated sites should hire remediation contractors on a fixed-price basis and should establish independent peer review panels to check progress toward specified remediation milestones.** Such steps are necessary to provide stronger incentives for federal remediation contractors to implement efficient, innovative solutions to contamination problems. When site complexities result in remediation costs that exceed the initial estimates, the peer review panel could verify that the cost increase is technically justified.

• **The EPA should review procedures for approving remediation technologies in its 10 regions and should develop guidelines for increasing the consistency and predictability of these procedures among regions and across programs; to the extent possible, state programs for contaminated site cleanup should follow these guidelines.** A consistent regulatory process that responds rapidly to approval requests is essential so that remediation technology developers can predict with reasonable certainty the steps that will be required for regulatory approval of their technology and how long they may have to wait before receiving their first job contract. While the process for remedy selection should be the same at each site, site managers must have the flexibility to consider any remediation technology that they believe will meet regulations at the lowest possible cost, provided the public has sufficient opportunity to voice concerns during the remedy selection process and to challenge the selected remedy.

• **Congress and the EPA should assess the arguments for and against establishing national standards for ground water and soil cleanup.** While some states are adopting state-wide cleanup standards, no national standards exist. Such standards might increase the predictability of the remediation process and consistency in the approaches used in the many remediation programs; predictability and consistency would benefit technology developers by providing them with a more certain end point for remediation. On the other hand, such standards might have the detrimental effect of decreasing flexibility in site remediation. The issue of whether national cleanup standards are advisable should be carefully considered.

• **The U.S. General Accounting Office should investigate the Massachusetts program for licensing site professionals to select remediation technologies on behalf of environmental regulators and should recommend whether such a program should be implemented nationally.** Such a program might help eliminate delays associated with regulatory approval steps.

• **The EPA should establish a national registry of contaminated sites similar to the Toxics Release Inventory and should make it publicly available on the Internet.** Such a registry would allow technology developers to assess the size and characteristics of different segments of the remediation market. It would also provide an incentive for companies to clean up sites quickly in order to remove them from the registry. Although there is political pressure to avoid including contaminated sites on registries because of the perceived stigma associated with owning a site on such a list, public disclosure of contaminated site information is essential for ensuring that accurate and complete information about the remediation market is widely available.

• **Federal agencies should continue to support and expand programs for testing innovative remediation technologies at federal facilities.** Providing opportunities for testing full-scale technology applications is essential for new technology ventures that need cost and performance data to provide to potential clients.

ADVANCING THE STATE OF THE PRACTICE OF GROUND WATER AND SOIL CLEANUP

Although considerable effort has been invested in ground water and soil cleanup, the technologies available for these cleanups are relatively rudimentary. Relatively effective and well-understood technologies are available for easily solved contamination problems—mobile and reactive contaminants in permeable and homogeneous geologic settings—but few technologies are available for treating recalcitrant contaminants in complex geologic formations. The greatest successes in remediation to date have been in the treatment of petroleum hydrocarbon fuels (gasoline, diesel, and jet fuel) because these are generally mobile and biologically reactive, but technologies for addressing other types of subsurface contamination problems are in short supply. Comparatively more technologies are available for treating contaminated soil than for treating contaminated ground water. While government agencies and others are investing considerable effort in remediation technology research, much more work in research, development, and field-scale application of remediation technologies is needed before ground water and soil contamination problems can be adequately solved.

The greatest challenge in remediation is in the location and cleanup of contaminant mass in the subsurface that can serve as a long-term source of ground water pollution and lead to the formation of extensive plumes of contamination. Plumes of ground water contamination generally originate from material existing in a nonaqueous phase (in other words, from masses of contaminants that initially are not dissolved in the water but that slowly dissolve when in contact with water). Sources of contamination may include organic solids, liquids, or vapors; inorganic sludges; compounds adsorbed on mineral surfaces; and compounds adsorbed in natural organic matter such as humus. Often, contaminant sources are difficult to locate and delineate. Once found, source material may be inaccessible, lying under structures, or at great depth, or in fractured rock. Because of the possibility of continual release of contaminants to ground water, partial source removal may not result in a proportional increase in ground water quality. The source may remain in place for a very long time because dissolution, while fast enough to create a potential hazard, may be too slow to result in rapid elimination of the source. Pumping treatment fluids to the region where sources are located may be very difficult due to hydrogeologic complexities. Added complexities arise during treatment of sources containing mixtures of contaminants because of the variable effects of treatment processes on different types of contaminants. Chapter 3 provides a detailed listing of specific research needed for improving the ability to clean up contaminant mass in ground water and soil.

Lack of information has contributed to the slow transfer of new ideas for remediation technologies from the laboratory to the field and from one site to another. Technology reports are often incomplete and lacking in critical scientific evaluation and peer review. Reliable cost data are also lacking. Moreover, much information on prior experiences with remediation technologies is proprietary. While several data bases on innovative technologies exist, none of these provides complete coverage of every application or test of every available remediation technology used in every remediation program. This lack of coordinated, high-quality information makes it difficult to compare technologies based on rational scientific evaluation.

Recommendations: Technology Information

Three types of activities are needed to improve the quality and availability of data on remediation technology performance. In addition to these activities, detailed research at the laboratory and field scales (see Chapter 3 for recommendations) is needed to increase the number of available remediation technologies and the efficiency of existing technologies.

• **The EPA, in collaboration with other stakeholders, should increase the scope and compatibility of data bases containing remediation technology performance information and should make these data bases available on the Internet, with a single World Wide Web page including links to all of the data bases.** Improvements in information collection, assessment, and dissemination are needed to speed development and commercialization of remediation technologies. While a single, centralized data base will likely be unwieldy and may not satisfy the diverse interests of various users, a goal for the EPA should be to develop comprehensive and electronically accessible data bases that can be readily distributed and manipulated by different contributors and users. A consistent framework for data entry and retrieval should be developed and used in all the data bases.

• **Government agencies, remediation consultants, and hazardous waste site owners should work to increase the sharing of information on remediation technology performance and costs.** Incentives should be developed to encourage submission of technology performance and cost data to the national data bases.

• **Government agencies, regulatory authorities, and professional organizations should undertake periodic, comprehensive peer review of innovative remediation technologies.** This type of activity will help define the state of the art, build consensus, and provide a standard for design and implementation of functional and cost-effective remediation technologies.

ESTABLISHING MEASURES OF SUCCESS FOR INNOVATIVE REMEDIATION TECHNOLOGIES

While many industries, such as the automotive and aerospace industries, have developed uniform standards for evaluating product performance, no such standards exist for ground water and soil cleanup technologies. Property owners responsible for site cleanup, citizen groups, state and federal regulators, and technology developers all may have different perspectives on how technologies should be evaluated and selected. There is currently no standardized mechanism for reconciling these differing expectations. Yet, to be widely applied, a technology must not only be a success in that it meets technical performance criteria, but it also must be accepted by these numerous stakeholders in site remediation. Any protocol used to test innovative remediation technologies must address common stakeholder expectations in some fashion if successful application is to follow.

Disagreements among stakeholders may arise due to many issues, but the critical disputes often center on the effectiveness of the technology in reducing health and environmental risks and the cost of the technology. Disputes over the level of risk reduction the technology must achieve arise because assessing the health and environmental risks of ground water and soil contamination is an uncertain process, and there is controversy over how to interpret results of risk assessments. Major uncertainty exists in determining accurate levels of exposure to contamination and the level of health or environmental damage caused by the contaminants. In evaluating remediation technologies, indirect quantitative criteria must substitute for a direct measure of the level of risk reduction the technology can achieve. The best measures for comparing the ability of different technologies to reduce health and environmental risks are the technology's ability to reduce contaminant mass, concentration, toxicity, and mobility because these criteria indicate the degree to which the technology can reduce the magnitude and duration of exposure to the contamination.

Disputes over costs of hazardous waste cleanup may arise because the affected public may want to "fix the contamination problem irrespective of costs" whereas site owners may wish to "manage the problem at the lowest possible cost." While there is no easy way to resolve this conflict, involving the public early in evaluating possible remedies for the site can minimize the acrimony. Anecdotal evidence from case studies examined in this report suggests that if the public were involved earlier in the decisionmaking process as a matter of routine, the universe of technologies taken under consideration might more routinely include innovative technologies.

Recommendations: Establishing Success Criteria

Increased attention to the concerns of all of the groups affected by hazardous waste sites is needed to streamline the process of remediation technology selection and to remove some of the obstacles to acceptance of innovative technologies. (See Chapter 4 for a detailed review of the factors of concern to stakeholders involved in site remediation.)

• **The EPA and state environmental regulators should amend their public participation programs and require that public involvement in contaminated site cleanup begin at the point of site discovery and investigation.** An informed public is better prepared to participate in the review of technology selection options and to consider innovative remediation technologies. Once site data are collected, the data should be made available at a convenient, accessible location of the public's choosing. While some members of the public desire short, factual data summaries, others may have expertise that equips them to review and evaluate the full studies, including laboratory analytical data and study protocol. To further assist the community, sources of toxicological and health information on contaminants of concern, as well as technical data collected from other sites where different technologies have been implemented and assessed, should also be provided.

• **Technology developers should consider the potential concerns of all stakeholders in remediation, including members of the public, when testing the performance of remediation technologies.** Even if a technology meets technical and commercial measures of success, strong opposition from the public or other stakeholders may make it undesirable.

• **Technology developers should report the effectiveness of their systems in reducing public health and environmental risks based on the technology's ability to reduce contaminant mass, concentration, mobility, and toxicity.** These measurable, technology-specific criteria are surrogates for environmental and health effects because they quantify the degree to which the technology can reduce exposure to the contamination. Technology developers should report the range of uncertainty in these measured values to allow for meaningful comparisons of risk reduction potential offered by different technologies.

• **Technology developers should specify the performance of their technology at the point of maximum effect and should indicate the distance of that point from the location where the technology is applied under some known or standardized flow or residence time conditions.** Depending on the technology and how it behaves in the field, the full effect of a technology in reducing risk may occur at some distance from the actual point of application. Specifying the point of maximum effect and its distance from the technology installation will improve technology comparisons.

TESTING INNOVATIVE TECHNOLOGIES

Just as there are no standard criteria for evaluating the success of innovative ground water and soil cleanup technologies, there are also no standard protocols for testing new product performance. This lack of protocols contributes to the difficulties that remediation technology developers face in trying to persuade potential clients that an innovative technology will work. The types of data collected for evaluating remediation technology performance vary widely and are typically determined by the preferences of the consultant responsible for selecting the technology, the client, and the regulators overseeing remediation at the contaminated site. Performance and cost data collected at one site are thus often insufficient for predicting how the technology will perform at another site. The Federal Remediation Technologies Roundtable has issued guidelines for collection and reporting of remediation technology performance data at federal sites, but no standard process exists for data collection and reporting at privately owned sites, and the degree to which the roundtable's guidelines are applied at federal facilities is unclear.

A variety of federal and state programs exists for evaluating remediation technology performance data, but these programs are not coordinated. For example, the state of California has different requirements for documenting technology performance than the Federal Remediation Technologies Roundtable, the Southern States Energy Board, and the Western Governors Association. The EPA has a national program, the Superfund Innovative Technologies Evaluation (SITE) program, for evaluating remediation technology performance data, but its scope and funding are limited. A technology developer may spend large sums testing a technology under one of the existing evaluation programs or according to one agency's procedures, only to learn that the data are not acceptable to potential clients or environmental regulators who are not specifically involved in the program under which the technology was evaluated.

As a result of the lack of standard procedures for collecting, reporting, and evaluating data on remediation technology performance, a great deal of money is spent on site-specific tests. Testing costs could be minimized if standard remediation technology performance data and a widely recognized national evaluation program were available. Some site-specific testing of remediation technologies will always be required prior to technology installation in most situations, but the requirements for site-specific testing would decrease if standard, verified data were easily available. Increasing standardization of data collection, reporting, and evaluation would also enable more accurate predictions of remediation technology performance at a new site. While the specific protocols used to test and evaluate a remediation technology will vary with the technology, common principles apply to all technologies, and standard types of performance data can be reported in a standard format for all types of technologies.

Recommendations: Technology Testing

Remediation technology developers, owners of contaminated sites, and environmental regulators all can take steps to increase the consistency in testing, reporting, and evaluation protocols for assessing remediation technology performance. (Chapter 5 provides a detailed description of general principles that should be followed in testing innovative remediation technologies and a set of guidelines for determining the amount of additional testing required to assess the performance of a remediation technology that has been used elsewhere.)

• **In proving performance of an innovative remediation technology, technology developers should provide data from field tests to answer the following two questions:**

1. Does the technology reduce risks posed by the soil or ground water contamination?
2. How does the technology work in reducing these risks? That is, what is the evidence proving that the technology was the cause of the observed risk reduction?

To answer the first question, the developer should provide two or more types of data, both leading to the conclusion that contaminant mass and concentration, and/or contaminant toxicity, and/or contaminant mobility decrease following application of the technology. To answer the second question, the developer should provide two or more types of evidence showing that the physical, chemical, and/or biological characteristics of the contaminated site change in ways that are consistent with the processes initiated by the technology.

• **In deciding how much site-specific testing to require before approving an innovative remediation technology, clients and environmental regulators should divide sites into the four categories shown in Figure ES-1: (I) highly treatable, (II) moderately difficult to treat, (III) difficult to treat, and (IV) extremely difficult to treat.** For category I sites, site-specific testing of innovative remediation technologies should be required only to develop design specifications; efficacy can be determined without testing, based on a review of fundamental principles of the remediation process, properties of the contaminated site, and prior experience with the technology. For category II sites, field pilot testing should be required to identify conditions that may limit the applicability of the technology to the site; testing requirements can be decreased as the data base of prior applications of the technology increases. For category III sites, laboratory and pilot tests will be necessary to prove efficacy and applicability of the technology at a specific site. For category IV sites, laboratory and pilot tests will be needed, and multiple pilot tests may be necessary to prove that the technology can perform under the full range of site conditions.

• **All tests of innovative remediation technology performance should include one or more experimental controls.** Controls (summarized in Chapter 5) are essential for establishing that observed changes in the zone targeted for remediation are due to the implemented technology. Failure to include appropri-

FIGURE ES-1 Treatability of contaminated sites and level of site-specific testing of remediation technologies required as a function of contaminant and geologic properties. Note that "H" indicates high and "L" indicates low volatility, reactivity, or solubility. (See Table ES-1 for a listing of the contaminant compounds.)

TABLE ES-1 Classes of Compounds Shown in Figure ES-1

Contaminant Class (as shown in Figure ES-1)	Volatility, Reactivity, and Solubility	Example Contaminants
A	HHL	Hydrocarbon fuels; benzene, toluene, ethylbenzene, and xylene
B	HLL	Trichloroethane, trichloroethylene, tetrachloroethylene
C	HHH	Acetone
D	LHH	Phenols, glycols
E	HLH	Methyl tertiary-butyl ether, tertiary butyl alcohol, methylene chloride
F	LHL	Naphthalene, small polycyclic aromatic hydrocarbons (PAHs), phthalates
G	LLH	Inorganic mixtures, metals of different chemistries
H	LLL	Polychlorinated biphenyls, pesticides, large PAHs

NOTES:

Volatility: High (H) > approximately 10 mm Hg; Low (L) < approximately 1 mm Hg

Reactivity: High indicates biodegradable, oxidizable compound; Low indicates recalcitrant compound

Solubility: High > approximately 10,000 mg/liter; Low < approximately 1,000 mg/liter

ate controls in the remediation technology performance testing protocol can lead to failure of the test to prove performance.

• **The EPA should establish a coordinated national program for testing and verifying the performance of new remediation technologies.** The program should be administered by the EPA and implemented either by EPA laboratories, a private testing organization, a professional association, or a nonprofit research institute. It should receive adequate funding to include the full range of ground water and soil remediation technologies and to test a wide variety of technologies each year. A successful test under the program should result in a guaranteed contract to use the technology at a federally owned contaminated site if the technology is cost competitive. The program should be coordinated with state agencies so that a technology verified under the program does not require additional state approvals.

• **Applications for remediation technology verification under the new verification program should include a summary sheet in standard format.** The summary sheet should include a description of the site at which the technology was tested, the evaluation methods used to prove technology performance, and the results of these tests. It should also include a table showing the types of data used to answer each of the two questions needed to prove technology performance.

• **Applications for remediation technology verification should specify the range of contaminant types and hydrogeologic conditions for which the technology is appropriate.** Separate performance data should be provided for each different major class of contaminant and hydrogeologic setting for which performance verification is being sought.

• **Data gathered from technology performance tests under the verification program should be entered in the coordinated national remediation technologies data bases recommended above.** Data should be included for technologies that were successfully verified and for those that failed the verification process.

• **Technology development partnerships involving government, industry, academia, and other interested stakeholders should be encouraged.** Such partnerships can leverage resources to speed innovative technologies through the pilot testing phase to commercial application.

COMPARING COSTS OF INNOVATIVE TECHNOLOGIES

One of the greatest challenges in selecting systems for ground water and soil cleanup is the development of reliable cost data. Comparing costs of different remediation technologies can be difficult in some situations and impossible in others, for several reasons. First, costs reported under a set of conditions at one site are difficult to extrapolate to other sites with different hydrogeologic and contaminant characteristics. Second, technology vendors may report costs using a variety of different metrics (such as dollars per volume treated, reduction in contaminants achieved, mobility reduction achieved, weight of contaminant removed, or surface area treated); these different metrics may not be comparable. Third, assumptions about what cost elements should be included in the cost estimates and what interest rates should be used to project long-term costs vary considerably. For example, cost estimating systems used by the federal government, such as the federal work breakdown structure, vary from those used by the private sector. Fourth, private-sector companies rarely compile cost information and release it for public use, so that it is often impossible to obtain actual cost data from completed projects.

Cost uncertainties are especially a problem when evaluating whether to use an innovative remediation technology. While remediation consultants can gauge the costs of conventional technologies based on their experience with prior applications of the technology, no historical record or only a very limited record exists for estimating costs of innovative remediation technologies. Uncertainties about costs add to the disincentives to select innovative remediation systems. The uncertainties about what the new technology will cost, combined with the technical uncertainty over whether it will perform as expected, may, in the client's view, outweigh the potential benefits of using the technology.

Recommendations: Comparing Costs

To improve the ability to compare costs of different remediation technologies and to extrapolate cost data from one site to another, a variety of strategies are needed to standardize current cost estimating and reporting procedures. (See Chapter 6 for a detailed critique of existing cost reporting procedures and a description of mechanisms for standardizing cost reporting.)

• **The EPA should convene a working group composed of representative problem owners (corporations and government agencies) and technology developers to develop and refine a standardized system of "template sites" for comparing the costs of ground water and soil remediation technologies.** The template sites should provide realistic models of contamination scenarios for use in developing cost comparisons. The EPA might convene the working group under the auspices of an established organization such as the Remediation Technologies Development Forum or the American Academy of Environmental Engineers. The working group should develop several templates to represent the range of conditions of contaminant depth, aquifer thickness, and aquifer permeability. Once the templates are developed and refined, federal agencies and private corporations should request that remediation technology vendors present cost data in the template format if the technology is to be evaluated for purchase. The templates can then be used to provide screening-level comparisons of remediation technologies designed to achieve the same level of public health and environmental protection. More detailed cost data, based on actual site conditions, would then need to be developed for the technologies that pass this first level of screening.

• **The Federal Remediation Technologies Roundtable should reevaluate the role of the work breakdown structure in standardizing federal remediation cost reporting and should document the system in a way that facilitates understanding by the private sector.** The federal work breakdown structure, a mechanism for tabulating costs of federal projects, may be too rigid in format to be appropriate for standardizing costs for the wide range of remediation needs and may not be an efficient tool for the private sector to use in developing cost estimates for new technologies. The role of the work breakdown structure should be reevaluated and a guidance manual prepared to help the private sector use this tool. The instruction manual should be advertised to remediation technology providers and users and should be available in an on-line version.

• **Costs of remediation technologies should be included in the coordinated national remediation technologies data bases recommended above and should always be reported as cost per unit volume of the contaminated ma-**

trix treated and as cost per weight of contaminant removed, treated, or contained. The starting concentration of the contaminant, amount of material cleaned up, and process rate should be provided along with these cost data because unit costs may change with the amount of contamination or contaminated material treated.

   • **Cost estimates should include one-time start-up costs as well as up-and-running costs.** Start-up costs include the costs of site preparation, equipment mobilization, pilot testing, permitting, and system design, yet frequently these are not included in cost estimates.

   • **Assumptions about discount rates and tax benefits should be clearly stated in estimates of present costs of a technology that operates over an extended time period.** In developing cost estimates for technology users, technology providers should tailor their assumptions about discount rates and taxes to the needs of the user, which vary widely.

   In summary, a combination of market incentives, research, and improved technology testing and evaluation strategies is needed to advance the capability to clean up ground water and soil at contaminated sites. Existing technologies have high costs and are inadequate for solving many types of contamination problems. If the United States is to protect the public health from risks associated with ground water and soil contamination, while avoiding needlessly exorbitant spending of taxpayer and corporate resources, then the federal government and others responsible for overseeing contaminated sites need to give high priority to the development of creative new solutions to site cleanup problems.

1

Challenges of Ground Water and Soil Cleanup

Over the past quarter century, the United States has placed a high priority on cleaning up sites where contaminants have leaked, spilled, or been disposed of in the soil and ground water. Anywhere from 300,000 to 400,000 contaminated sites are scheduled for cleanup in the coming decades, at an estimated total cost as high as $500 billion to $1 trillion (National Research Council, 1994; Russell et al., 1991). The Office of Management and Budget estimates that the costs of remediation at contaminated sites on property owned by the Departments of Defense, Energy, Interior, and Agriculture and the National Aeronautics and Space Administration will total between $234 and $389 billion over the next 75 years (Federal Facilities Policy Group, 1995). National spending on waste site remediation totaled an estimated $9 billion in 1996 alone, as shown in Figure 1-1.

As cleanup at waste sites has proceeded, it has become increasingly recognized that despite the billions of dollars invested, conventional remediation technologies, especially for sites with contaminated ground water, are inadequate. For example, a 1994 National Research Council (NRC) study of conventional ground water cleanup systems at 77 contaminated sites determined that ground water cleanup goals had been achieved at only 8 of the sites and that full achievement of cleanup goals was highly unlikely with the in-place technologies at 34 of the 77 sites (NRC, 1994; MacDonald and Kavanaugh, 1994, 1995). A 1995 review by the Congressional Budget Office found that using nonconven-tional methods for waste site investigation and cleanup could cut costs by 50 percent or more (CBO, 1995). Based on such findings, it is clear that new technologies are needed to restore the nation's contaminated sites.

The limitations of conventional ground water cleanup systems are now well recognized by regulators, consultants, engineers, and others involved in waste

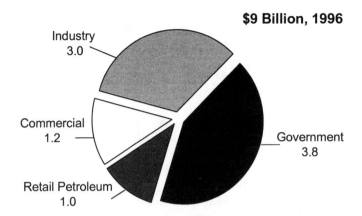

FIGURE 1-1   Estimated spending on environmental remediation (in billions of dollars) in 1996 in various sectors of the U.S. economy. Total spending was estimated at $9 billion. SOURCE: Adapted from information in Environmental Business International, 1995.

site remediation. Indeed, federal and state agencies and many private industries have launched numerous initiatives to promote the development and use of innovative remediation technologies. These initiatives range from issuance of an official EPA policy titled "Initiatives to Promote Innovative Technology in Waste Management Programs" (Laws, 1996), to development of the Ground Water Remediation Technologies Analysis Center (which provides on-line information on new technologies) (GWRTAC, 1995), to establishment of government/industry partnerships such as the Remediation Technologies Development Forum. The EPA has an office, the Technology Innovation Office, dedicated to finding ways to promote use of innovative remediation technologies. These and other initiatives have led to increased research on innovative remediation technologies.

While there has been major progress in research and increased use of new technologies in some situations, use of new ground water cleanup technologies at major contaminated sites is still limited. For example, as of 1996, conventional pump-and-treat methods were being used for ground water cleanup at 93 percent of Superfund sites (EPA, 1996). The reasons for the limited use of new ground water cleanup technologies are complex. They include regulatory programs that inhibit market development, lack of consistent data on technology cost and performance, and the uniqueness of each contaminated site.

The lack of commercially available technologies that can restore contaminated ground water at reasonable cost has led to increasing pressure to limit waste cleanups to sites that pose immediate risks to human health, rather than applying costly and potentially ineffective conventional cleanup systems. The American Society for Testing and Materials (ASTM) in 1995 issued a standard entitled "Standard Guide for Risk-Based Correction Action Applied at Petroleum-Release Sites" (known as RBCA) that outlines a procedure for limiting the cleanup of underground storage tank sites to those posing immediate risks (ASTM, 1995). RBCA is a process for determining site-specific risk factors and setting site-specific cleanup goals. The standard is controversial because of uncertainty in the risk assessment methods it employs, but many states are implementing it in cleanups of petroleum contamination from underground storage tanks. In addition, many organizations are now lobbying to implement a similar ASTM RBCA standard for chemical release sites, which is under development, at major contaminated sites regulated under federal programs. If RBCA were widely applied at all types of contaminated sites, a large fraction of sites currently slated for remediation would not be actively cleaned up (Begley, 1996), and those that are would be cleaned up to less stringent standards.

While the lack of cost-effective, commercially available remediation technologies has led to increased use of RBCA as a means for limiting site cleanups, the development of effective technology can cause a counter trend. When technology becomes available to address contamination at affordable costs, pressure to apply the technology on a widespread basis will increase. When contaminants are left in place, those responsible for the contaminated site must bear the cost of continued liability should contamination escape from the site into surrounding communities. Predicting the potential for such contaminant migration off site is subject to significant uncertainties, so that the full costs of this long-term liability are difficult to calculate. Further, maintaining the site to prevent exposure to the contamination may involve long-term costs. Costs are also associated with decreased property values and difficulty in selling property when significant contamination remains in place. Finally, leaving contamination in place is unacceptable to members of some communities near contaminated sites. Affordable remediation technologies that can remove the bulk of contaminant mass from the subsurface at contaminated sites would reduce the long-term risks, liabilities, and costs associated with these sites.

This report focuses on how to harness market forces to stimulate development of new, affordable remediation technologies and how to standardize testing, evaluation, and cost comparison of innovative remediation technologies. Standardizing technology testing and data collection is an important step in commercializing innovative remediation technologies and in reducing costs because current data sets are often inadequate for extrapolating data from one site to the design of a cleanup system at another site. As explained in Chapter 2 of this report, a necessary prerequisite to the establishment of standardized testing programs for remediation technologies is assurance that strong market forces are in place to stimulate demand for new technologies. No amount of government promotion of technology testing will be fully effective if the market demand for innovative technologies is lacking.

This report is the culmination of a two-and-a-half-year study by the NRC's Committee on Innovative Remediation Technologies. The committee was appointed by the NRC in 1994 to develop testing and performance standards for subsurface cleanup technologies and to examine other issues related to commercialization of these technologies. The members of the committee, who are the authors of this report, included environmental consultants, environmental researchers from academia, and experts in environmental policy, patent law, technology financing, and public opinion. In conducting its study, the committee consulted with a wide range of stakeholders involved in the testing of subsurface cleanup technologies, including federal and state regulators, industry groups, heads of start-up technology companies, and venture capitalists.

This chapter provides an overview of the sources of ground water and soil contamination, the limitations of conventional remediation technologies, and the frequency of use of innovative remediation technologies. Chapter 2 assesses the remediation technology market and recommends market-based approaches for strengthening it. Chapter 3 defines the current state of the practice in ground water and soil cleanup, identifying areas where innovation is needed. Chapter 4 outlines benchmark criteria for evaluating ground water and soil cleanup technologies to satisfy the concerns of all stakeholders. Chapter 5 recommends testing strategies for evaluating technology performance. Chapter 6 describes how to compare the costs of alternative technologies.

SOURCES OF GROUND WATER AND SOIL CONTAMINATION

Accidental spills, routine washing and rinsing of machinery and chemical storage tanks, leaks in industrial waste pits and municipal and industrial landfills, and a variety of other human activities (see Table 1-1 and Figure 1-2) can release contaminants to soil and ground water (see Box 1-1). The most common types of contaminants found at waste sites are chlorinated solvents, petroleum hydrocarbons, and metals (NRC, 1994). Chlorinated solvents, such as trichloroethylene and perchloroethylene (PCE), are used for purposes ranging from dry cleaning of consumer goods to degreasing of industrial manufacturing equipment and cleaning of military aircraft. Petroleum hydrocarbons commonly found in ground water include the components of gasoline (benzene, toluene, ethylbenzene, and xylene, together known as BTEX), as well as other fuels. Because of the widespread use of both chlorinated solvents and petroleum hydrocarbons, it is not surprising that they are found in the ground water at hundreds of thousands of contaminated sites across the country. Other contaminants found in ground water and soil at many sites are polycyclic aromatic hydrocarbons, created from combustion, coal coking and processing, petroleum refining, and wood treating operations; polychlorinated biphenyls (PCBs), once widely used in electrical transformers and capacitors and for a variety of other industrial purposes; pesticides, used for agriculture; metals, from metal plating and smelting operations, mines, and other industrial activities; and radioactive compounds, used in the manufacture of nuclear weapons at Department of Energy (DOE) facilities.

TABLE 1-1   Ground Water Contamination Sources

Sources designed to discharge substances
Subsurface percolation (e.g., septic tanks and
  cesspools)
Injection wells
  Hazardous waste
  Nonhazardous waste (e.g., brine disposal
    and drainage)
  Nonwaste (e.g., enhanced oil recovery,
    artificial recharge, solution mining, and
    in situ mining)
Land application
  Wastewater (e.g., spray irrigation)
  Wastewater byproducts (e.g., sludge)
  Hazardous waste

**Sources designed to store, treat, and/or
dispose of substances; discharge through
unplanned release**
Landfills
  Industrial hazardous waste
  Industrial nonhazardous waste
  Municipal sanitary
Open dumps, including illegal dumping
Residential (or local) disposal
Surface impoundments
Waste tailings
Waste piles
Materials stockpiles
Graveyards
Animal burial sites
Above-ground storage tanks
Underground storage tanks
Containers
Open burning and detonation sites
Radioactive disposal sites

**Sources designed to retain substances during
transport or transmission**
Pipelines
Material transport and transfer operations

**Sources discharging substances as
consequences of other planned activities**
Irrigation practices (e.g., return flow)
Pesticide applications
Fertilizer applications
Animal feeding operations
De-icing salts applications
Urban runoff
Percolation of atmospheric pollutants
Mining and mine drainage

**Sources providing pollution conduits or
inducing discharge through altered flow
patterns**
Production wells
  Oil (and gas) wells
  Geothermal and heat recovery wells
  Water supply wells
Other wells
  Monitoring wells
  Exploration wells
Construction excavation
Drains

**Naturally occurring sources, with discharge
created and/or exacerbated by human
activity**
Ground water–surface water interactions
Natural leaching
Salt water intrusion/brackish water upconing
  (or intrusion of other poor-quality natural
  water)

SOURCE: Adapted from Reichard et al. (1990).

As shown in Table 1-1, some contaminants are released directly to ground water, for example in water injection wells, while others are released to the soil. When released to the soil, contaminants will migrate through the soil and may contaminate the underlying ground water (see Box 1-2). Some contaminants may dissolve in the ground water as it percolates through the soil. Others may dissolve in the gases contained in soil pores and spread before dissolving in the ground water. Contaminants also may be transported as a separate, nonaqueous-phase

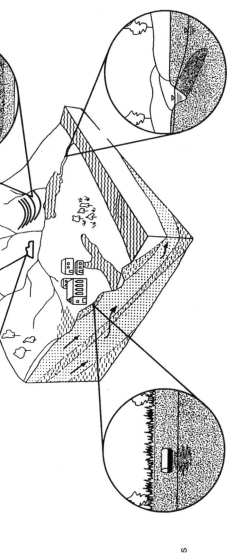

POINT SOURCES

• Underground storage facilities
• Landfills and hazadous waste disposal sites
• Surface impoundments
• Illegal disposal of waste or toxic chemicals
• Industrial areas
• Septic tanks
• Transportation spills and accidents
• Injection wells and boreholes
• Urban storm-water runoff into sinks
• Unplugged oil and gas wells

LINE SOURCES

• Underground pipelines
• Sewage canals
• Surface water (rivers)
• Roads
• Railway track

AREAL SOURCES

• Urban areas
• Industrial ares
• Agricultural areas
• Mining waste sites
• Atmospheric components

FIGURE 1-2 Classification of ground water contamination sources according to source geometry. SOURCE: Reprinted, with permission, from Spitz and Moreno (1996). © 1996 by John Wiley & Sons.

BOX 1-1
The Underground Environment

The underground environment consists of layers of granular materials (such as sand and gravel), clay, and solid rock. Ground water flows through the pores and fractures in these materials, in formations known as aquifers. There are two kinds of aquifers: consolidated and unconsolidated (see Figures 1-3 and 1-4). Consolidated aquifers consist of essentially solid rock permeated with cracks and crevices through which water flows. Unconsolidated aquifers consist of uncemented granular materials; water and other fluids flow through the pore spaces among these materials. Below the water table, all of the pores and crevices in an aquifer are saturated with water. This region is technically known as the "saturated" zone. Above the water table, the pores and crevices are only partially filled with water. This region is known as the "unsaturated" or "vadose" zone.

Geologic processes can produce aquifers with highly variable (heterogeneous) hydraulic and geochemical properties. For example, sand and gravel aquifers may contain lenses of clay. Even in relatively homogeneous aquifers, the grain size of aquifer materials may vary with location across a small area. Because of the nonuniformity of aquifer formations, the flow of water and other liquids through the subsurface can be difficult to predict.

---

liquid (known as a NAPL) that is immiscible in water and therefore travels separately from the water. Other contaminants can sorb to mobile colloidal particles or form complexes with molecules of natural organic matter present in the water and be transported with these particles and complexes.

The fate of contaminants once released to the soil or ground water is extremely difficult to predict for a variety of reasons. Contaminated fluids (water, gas, and NAPLs) will flow preferentially through soil pathways offering the least resistance, and the locations of these pathways may be very difficult to determine. Contaminants also may sorb to the soil, or, in the case of metals, precipitate. NAPL contaminants may become entrapped in soil pores, leaving residual-phase contamination. Once the soil pores are saturated with NAPLs, the remaining NAPL will migrate downward to the water table. If the NAPL is less dense than water, it may form a pool at the surface of the water table. If the NAPL is more dense than water, it will continue its downward migration—in some cases in the form of narrow, viscous "fingers" that are extremely difficult to locate—until it encounters an impermeable barrier (NRC, 1994). Under each of these circum-

Coal tar recovered from a ground water monitoring well. This dense nonaqueous-phase liquid can contaminate water with polycyclic aromatic hydrocarbons and other aromatic contaminants. The liquid is extremely difficult to locate and remove once it migrates into the subsurface. Courtesy of Richard Luthy, Carnegie Mellon University.

stances, the contaminants will leave a reservoir that will serve as a long-term source of ground water contamination.

While contaminant source areas may be small and present no immediate hazard to human health or the environment, contaminants from these source areas will dissolve very slowly in the passing ground water, forming a plume and spreading. The plume can migrate large distances and contaminate drinking water wells, wetlands, and receiving waters. The size and location of the plume depend on the location of the contaminant sources, the path of natural ground water flow, and the various subsurface mechanisms that can entrap or transform the contaminant. The speed at which the plume will move depends on the rate of ground water flow and on contaminant retention and transformation mechanisms. Generally, the average ground water flow rate will be the maximum possible average rate of plume movement. Ground water flow rates vary widely from site to site depending on local hydrogeology, with values ranging from less than 1 mm per day to more than 1 m per day. Other processes occurring in the subsurface (see Boxes 1-3 and 1-4) cause the contaminant to move more slowly than the ground water.

As an example of the complexity of contaminant flow paths, Figure 1-5 shows the migration of PCE at a hypothetical site. The black areas contain undissolved PCE that has migrated separately from the ground water. The shaded por-

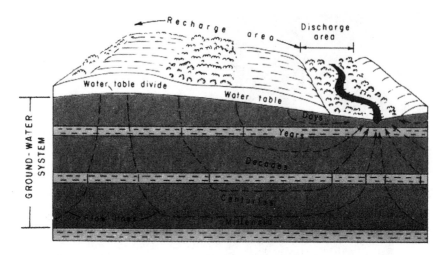

FIGURE 1-3   Simplified schematic of ground water flow in an unconsolidated aquifer. The flow lines indicate travel times to various parts of the subsurface, with longer travel times indicated by flow lines reaching deeper into the subsurface. SOURCE: Heath (1983) as reprinted in NRC, 1994.

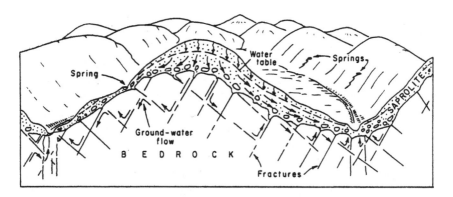

FIGURE 1-4   Simplified schematic of ground water flow in a consolidated aquifer. As the flow lines indicate, the direction of ground water flow in such aquifers depends on the locations of the fractures and thus is often tortuous and difficult to predict. SOURCE: From Heath (1980), as reprinted in NRC, 1994.

BOX 1-2
Contaminant Transport Mechanisms

Contaminants may move underground by one or a combination of several mechanisms, depending on their properties:

1. Vapor-phase transport: Vapors of volatile contaminants may spread through the pore spaces in the soil above the water table and then either dissolve in water in soil pore spaces or in infiltrating rain water. The volatility of contaminants, and thus the extent to which they will migrate in the vapor phase, varies by many orders of magnitude.

2. Aqueous-phase transport: Contaminants may dissolve in and be transported with the flowing ground water. The rate of dissolution depends on contaminant solubility, which varies among contaminants by many orders of magnitude; the extent of contaminant contact with water; and contaminant reactions with solids in the aquifer.

3. Nonaqueous-phase liquid (NAPL) transport: Many contaminants, including chlorinated solvents and petroleum products, enter the subsurface in the form of an oily liquid, known as a NAPL. NAPLs do not mix readily with water and therefore flow separately from ground water. If the NAPL is more dense than water (known as a DNAPL), it will tend to sink once it reaches the water table. If the liquid is less dense than water (known as an LNAPL), it will tend to float on the water table.

4. Facilitated transport: Contaminants may sorb to mobile colloidal particles or be incorporated into large complexes of natural organic matter and be transported with these particles or complexes in the flowing ground water. Contaminants associated with colloidal particles and organic complexes can travel with the ground water at rates much faster than would be predicted based upon contaminant transport models that neglect to consider these reactions. Such reactions are especially significant for metals and radionuclides; these contaminants generally have limited solubility over the pH range encountered in most ground waters, but sorption and complexation reactions can greatly increase the quantity of contaminant in the water.

As an example of these transport pathways, Figure 1-5 illustrates the possible fate of perchloroethylene in an aquifer consisting of strata of sand and fractured clay.

BOX 1-3
Contaminant Retention Mechanisms

A variety of physical and chemical processes can retain contaminants in the subsurface. While the interactions governing retention vary with contaminant and site characteristics, the effect on contaminant transport in all cases is the same: a retardation in the average rate of movement of the contaminant with respect to the ambient ground water flow. Key mechanisms for retention include the following:

• *Sorption and ion exchange:* Contaminants may sorb to solid materials in the subsurface. Contaminants such as heavy metals, polycyclic aromatic hydrocarbons, PCBs, and some pesticides have a strong tendency to sorb to soil under chemical conditions commonly found in the subsurface.

• *NAPL entrapment:* As illustrated in the example for PCE, small globules of NAPLs can become trapped in porous materials by capillary forces. The amount of entrapped compound is quantified technically as "residual saturation," the ratio of the entrapped volume of NAPL to the total pore volume.

• *Diffusion into micropores:* Dissolved contaminants may migrate by molecular diffusion into tiny micropores within aggregates of geologic materials.

• *Entrapment in immobile zones:* Contaminants may migrate into geologic zones where the ground water flow rate is very slow, essentially zero.

• *Precipitation:* Depending on the pH and other chemical characteristics of the ground water, metal contaminants may precipitate, forming immobile solids.

---

tions of the figure show the movement of the plume of dissolved PCE from the source areas containing undissolved PCE.

TYPES OF CONTAMINATED SITES

In general, hazardous waste sites can be grouped into the following seven categories (NRC, 1994):

1. closed or abandoned waste sites designated for cleanup under the Comprehensive Environmental Response, Compensation, and Liability Act (CERCLA), commonly known as Superfund;

BOX 1-4
Contaminant Transformation Mechanisms

Within the subsurface, a variety of biological and chemical reactions can degrade contaminants to harmless end products or transform them into other hazardous compounds. Naturally occurring microorganisms in the subsurface can use contaminants as sources of food and energy when there is a sufficient supply of oxygen or other substances that can serve as electron acceptors, suitable pH, and a sufficient quantity of nutrients. The microorganisms convert the contaminants to harmless end products such as carbon dioxide, methane, hydrogen sulfide, nitrogen gas, and water. For example, in ground water systems containing sufficient oxygen, microbes can degrade gasoline to carbon dioxide and water relatively easily. Under the right geochemical conditions, microorganisms can adapt to degrade many types of organic compounds, but this adaptation may require a long time period (NRC, 1993; MacDonald and Rittmann, 1993).

Chemical reactions in the subsurface can also degrade or transform contaminants. For example, under the right geochemical conditions, metals can precipitate, forming relatively immobile solids.

Predicting the degree to which microbial or chemical reactions will transform contaminants at a particular location is complicated by the non-uniform distribution of microorganisms in the subsurface and by the high variability of subsurface geochemistry. The subsurface is composed of a large number of microenvironments. For example, pores may be large or small, open or closed at the top, and located in sedimentary material with varying mineral composition (Chapelle, 1992). This diversity of environments creates diversity in the types of microbial communities and possible biochemical transformations that may be achieved at any one location in the subsurface (Chapelle, 1992).

---

2. active waste management facilities regulated under the Resource Conservation and Recovery Act (RCRA);

3. leaking underground storage tanks at gas stations and other facilities, for which cleanup is required under a special section of RCRA;

4. Department of Defense (DOD) facilities, which must be cleaned up to meet the requirements of the Federal Facilities Compliance Act and to prepare the land for sale when the facility is closed;

5. Department of Energy (DOE) facilities, which must also be cleaned up under the Federal Facilities Compliance Act;

6. federal facilities managed by agencies other than DOD and DOE, which also require cleanup under the Federal Facilities Compliance Act; and

7. sites governed by state hazardous waste programs.

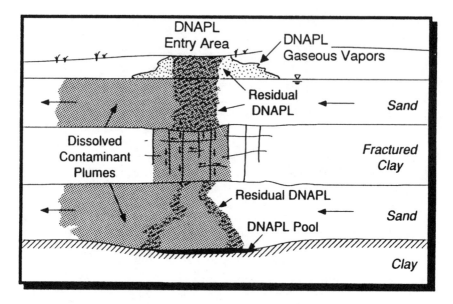

FIGURE 1-5   Possible fate of PCE in an aquifer consisting of sand and fractured clay. Some of the PCE volatilizes, some is entrapped as a residual in soil pores, some migrates into the fractures, and some pools on the clay layer. PCE from all of these sources dissolves in the flowing ground water. SOURCE: Reprinted, with permission, from Cohen and Mercer (1993). © 1993 by C. K. Smoley.

Table 1-2 shows estimates of the number of sites in each of the above categories. The estimates vary somewhat depending on the source of the evaluation. In general, the estimated total number of contaminated sites is in the range of 300,000 to 400,000. Most of these sites are contaminated as a result of leaks in underground storage tanks.

An undetermined number of sites governed under CERCLA, RCRA, leaking underground storage tank, and state cleanup regulations are on idle industrial property, known as "brownfields," that state and local governments would like to redevelop (GAO, 1995; OTA, 1995). For example, Illinois state officials estimate that 5,000 brownfield sites exist in the state (GAO, 1995). In Chicago, 18 percent of industrial land is currently idled (GAO, 1995). The limited effectiveness of subsurface cleanup technologies has contributed to the difficulty of redeveloping these properties. Lenders have hesitated to provide loans for purchasing or developing brownfield sites where contamination remains in place because of concerns that they will be held liable for cleanup at some future date, that the value of the property will be depressed due to contamination, or that the site owners will be unable to repay the loan if they incur major expenses in trying to clean up the

TABLE 1-2   Number of Hazardous Waste Sites Where Ground Water May Be Contaminated

Site Category	Source of Estimate		
	EPA, 1993	Russell et al., 1991	Office of Technology Assessment, 1989
CERCLA National Priorities List	2,000	3,000	10,000
RCRA corrective action	1,500-3,500	NA	2,000-5,000
Leaking underground storage tanks	295,000	365,000	300,000-400,000
Department of Defense	7,300 (at 1,800 installations)	7,300	8,139
Department of Energy	4,000 (at 110 installations)	NA	1,700
Other federal facilities	350	NA	1,000
State sites	20,000	30,000	40,000
Approximate total	330,000	NA	360,000-470,000

NOTE: The numbers presented in this table are estimates, not precise counts. In addition, at some of these sites, ground water may not be contaminated. For example, the EPA (1993) estimates that ground water is contaminated at 80 percent of CERCLA National Priorities List sites. There is also some overlap in site categories. For example, 7 percent of RCRA sites are federal facilities, and 23 DOE sites are on the CERCLA National Priorities List (EPA, 1993). NA indicates that an estimate comparable to the other estimates is not available from this source. SOURCE: NRC, 1994.

site. State and local governments around the nation are currently creating programs to encourage redevelopment of brownfield sites, and these brownfield programs have become major drivers in the remediation marketplace, along with the cleanup programs listed in Table 1-2.

It is important to note that the complexity of waste sites varies enormously depending on the source of the contamination and the geologic conditions at the site. Most of the sites shown in Table 1-2 are leaking underground storage tank sites. If the contamination is from a single tank at a gas station, the site will be relatively easy to clean up, especially if it is located in an area with relatively homogeneous geology. On the other hand, contaminated sites at major DOD and DOE installations, as well as at many industrial facilities, may contain complex mixtures of chlorinated solvents, fuels, metals, and, at DOE facilities, radioactive substances. These contaminants may have leaked, spilled, or been disposed of into the ground water over several decades, creating contamination problems that, in turn, will require decades to clean up.

LIMITATIONS OF CONVENTIONAL
REMEDIATION TECHNOLOGIES

As is now widely recognized, conventional methods for cleaning up ground water and soil at hazardous waste sites have met with limited success.

Conventional technologies for cleaning contaminated ground water are based on the principle that if enough water is pumped from the site, the contaminants will eventually be flushed out. These conventional technologies are known as "pump-and-treat" systems (see Figure 1-6) because they pump water from the site and treat it to remove the contamination. For several reasons, the flushing process employed by pump-and-treat systems has limited effectiveness, especially for cleaning up undissolved sources of contamination beneath the water table. Key contaminant and subsurface properties that interfere with flushing include the following (NRC, 1994; MacDonald and Kavanaugh, 1994, 1995):

• *Immiscibility of contaminants with water:* Many contaminants are extremely difficult to flush from the subsurface because of their relatively low solubility in water.
• *Diffusion of contaminants into micropores and zones with limited water mobility:* The microscopic pores and zones with limited water mobility into which

Excavation of soil at a contaminated site.   Courtesy of Fluor Daniel GTI.

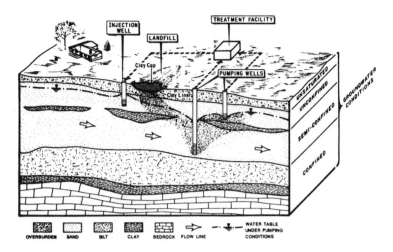

FIGURE 1-6 Conventional pump-and-treat system for cleanup of contaminated ground water. SOURCE: From Mercer et al. (1990) as reprinted in NRC, 1994.

contaminants may diffuse are extremely difficult to flush with water because of their small size and inaccessibility.

• *Sorption of contaminants to subsurface materials:* Flushing out contaminants that have sorbed to underground soils is a very slow process because of the slow rate of desorption.

• *Heterogeneity of the subsurface:* Prediction methods for determining the routes of travel of contaminants and of water used to flush out contaminants are not always accurate because of the heterogeneous nature of the subsurface.

Because of the difficulty of flushing contaminants from the subsurface, the NRC concluded in its 1994 study that pump-and-treat systems would be unable to fully restore many types of contaminated sites (see Box 1-5).

Historically, the conventional approach to soil cleanup has been to incinerate the contaminated soil on site or off site, to solidify it in place with cementing agents, or to excavate it and dispose of it in a hazardous waste landfill. The public often objects to incineration because of the air pollution it can create, and cleanup of many Superfund sites has been halted because of such objections. One of many such examples is the Baird & McGuire Superfund site in Massachusetts, where citizens formed a lobbying group, Citizens Opposed to Polluting the Environment, to block installation of an incinerator (MacDonald, 1994). Solidification technologies and excavation, while less controversial, are limited in that they do not clean up the contamination but simply immobilize it or move it elsewhere. All of these traditional remedies, especially incineration and excavation involving transport of the excavated materials, are costly. For example, cleanup of PCB-

---

BOX 1-5
Performance of Conventional Pump-and-Treat Systems

In its 1994 study, the NRC developed a scale of 1 through 4, shown in Table 1-3, for categorizing sites according to their difficulty of cleanup with conventional pump-and-treat systems. As shown in the table, the categories are based on the hydrogeology of the site or portion of the site and the chemistry of the contaminants. The 1994 study concluded that while cleanup of sites in category 1 (those with relatively simple geology and contaminant chemistry) to drinking water standards should be feasible with conventional pump-and-treat systems, cleanup of sites in category 4 is unlikely. The study determined that cleanup of sites in categories 2 and 3 may be feasible in some situations but is subject to uncertainties that may prevent the achievement of cleanup goals, especially for sites in category 3.

The study included a review of pump-and-treat systems operating at 77 sites chosen based on the availability of information. The distribution of sites was not representative of the distribution of all types of waste sites nationwide because fewer than 10 percent of the 77 sites were service stations, while at least 80 percent of the contaminated sites nationwide are underground storage tank sites (see Table 1-2), and many of these are service stations. However, with the exception of the service stations, the 77 sites were more representative of the types of sites regulated under Superfund and RCRA.

Of the 77 sites reviewed in the study,

* 2 were in category 1, and cleanup goals had been achieved at 1 of these sites;
* 14 were in category 2, and cleanup goals had been achieved at 4 of these sites;
* 29 were in category 3, and goals had been achieved at 3 of these sites; and
* 42 were in category 4, and cleanup goals were achieved at none of these sites.

---

contaminated soil using conventional methods can cost as much as $2,000 per ton of soil.

USE OF INNOVATIVE REMEDIATION TECHNOLOGIES

During the 1990s, as the limitations of conventional subsurface remediation technologies have become increasingly clear, innovative technologies have become increasingly common in the cleanup of contaminated soil and of leaking

TABLE 1-3   Relative Ease of Cleaning Up Contaminated Aquifers as a Function of Contaminant Chemistry and Hydrogeology

Hydrogeology	Contaminant Chemistry					
	Mobile, Dissolved (degrades/ volatilizes)	Mobile, Dissolved	Strongly Sorbed, Dissolved[a] (degrades/ volatilizes)	Strongly Sorbed, Dissolved[a]	Separate Phase LNAPL	Separate Phase DNAPL
Homogeneous, single layer	1[b]	1-2	2	2-3	2-3	4
Homogeneous, multiple layers	1	1-2	2	2-3	2-3	3
Heterogeneous, single layer	2	2	3	3	3	4
Heterogeneous, multiple layers	2	2	3	3	3	4
Fractured	3	3	3	3	4	4

[a]"Strongly sorbed" generally indicates contaminants for which the retardation coefficient is greater than 10. A retardation coefficient of 10 indicates that at any given time, 10 percent of the contaminant is dissolved in the water and 90 percent is sorbed to the aquifer solids.

[b]Relative ease of cleanup, where 1 is easiest and 4 is most difficult.

SOURCE: NRC, 1994.

underground storage tanks containing petroleum products. However, use of innovative technologies is still very rare for cleaning up ground water at major contaminated sites regulated by the Superfund and RCRA programs.

Figures 1-7 and 1-8 show the types of technologies used to clean up contaminated soil at Superfund and underground storage tank sites, respectively. As shown in Figure 1-7, innovative technologies have been selected for cleaning up contaminated soil, sludge, and sediments at 43 percent of Superfund sites. However, the number of innovative technologies in use at these sites is limited. Two technologies, soil vapor extraction and thermal desorption, accounted for more than half of the innovative technologies selected. As shown in Figure 1-8, innovative approaches were chosen at approximately 66 percent of underground storage tank sites. However, landfilling is still the predominant remedy for contaminated soil at these sites.

Figures 1-9 and 1-10 show the types of technologies used to clean up contaminated ground water at Superfund and underground storage tank sites, respectively. As shown, conventional pump-and-treat systems are the chosen remedy at 93 percent of Superfund sites with contaminated ground water; in situ treatment remedies not involving pump-and-treat systems are used at fewer than 1 percent

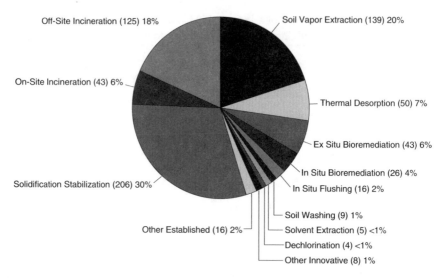

FIGURE 1-7   Types of technologies used to clean up contaminated soil at Superfund sites. Data for off-site incineration, solidification/stabilization, and other established technologies are based on records of decision for fiscal years 1982 through 1993. Data for innovative technologies and on-site incineration are based on anticipated design and construction activities as of August 1996. A site may use more than one technology. ( ) indicates the number of times this technology was selected or used. "Other" established technologies are soil aeration, open detonation, and chemical neutralization. "Other" innovative technologies are hot air injection, physical separation, contained recovery of oily wastes (CROW™), cyanide oxidation, vitrification, and plasma high temperature metals recovery. SOURCE: Adapted from EPA, 1996.

of the sites. At underground storage tank sites, innovative technologies are being used at approximately 43 percent of sites where active remedies (other than intrinsic remediation) have been selected. The greater use of innovative ground water cleanup technologies at underground storage tank sites in comparison to Superfund sites is a function of the relative simplicity of cleaning up these sites in comparison to Superfund sites and the greater regulatory flexibility of the underground storage tank program. Leaking underground storage tanks typically contain petroleum hydrocarbons, which are generally easier to clean up than other types of contaminants (see Chapter 3). In addition, these sites are relatively small in comparison to Superfund sites. Finally, underground storage tank cleanups are regulated by state agencies, and typically there is minimal regulatory oversight in technology selection, allowing greater freedom to choose different types of technologies (see Chapter 2).

Comprehensive data such as are available for the Superfund and underground

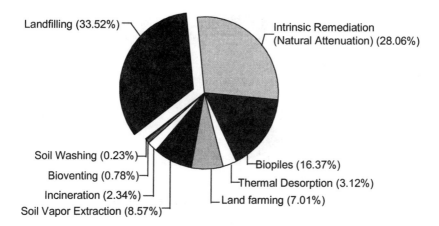

Landfilling (33.52%)

Intrinsic Remediation
(Natural Attenuation) (28.06%)

Soil Washing (0.23%)
Bioventing (0.78%)
Incineration (2.34%)
Soil Vapor Extraction (8.57%)

Biopiles (16.37%)
Thermal Desorption (3.12%)
Land farming (7.01%)

FIGURE 1-8   Types of technologies used to clean up contaminated soil at underground storage tank sites. The total number of sites where soil cleanup is under way is approximately 103,000. SOURCE: Adapted from Tremblay et al., 1995.

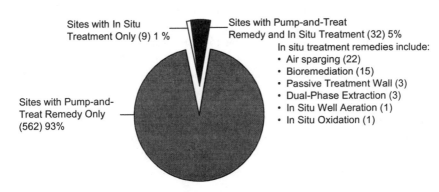

Sites with In Situ
Treatment Only (9) 1 %

Sites with Pump-and-Treat
Remedy and In Situ Treatment (32) 5%
In situ treatment remedies include:
• Air sparging (22)
• Bioremediation (15)
• Passive Treatment Wall (3)
• Dual-Phase Extraction (3)
• In Situ Well Aeration (1)
• In Situ Oxidation (1)

Sites with Pump-and-
Treat Remedy Only
(562) 93%

FIGURE 1-9   Types of technologies used to clean up contaminated ground water at Superfund sites. Pump-and-treat remedy data are based on records of decision for fiscal years 1982 through 1995; in situ treatment data are based on anticipated design and construction activities for August 1996. The total number of sites with remedies for contaminated ground water is 603. The total number of in situ treatment remedies exceeds the total number of sites at which treatment remedies are being implemented because more than one technology is being employed at some sites. SOURCE: Adapted from EPA, 1996.

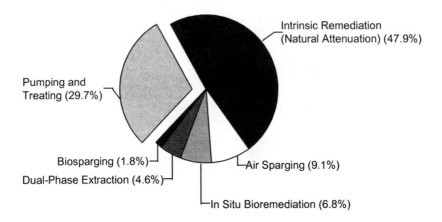

FIGURE 1-10   Types of technologies used to clean up contaminated ground water at underground storage tank sites. The total number of sites where ground water cleanup is under way is approximately 19,200. SOURCE: Adapted from Tremblay et al., 1995.

storage tank programs are not available for the RCRA program or for federal facilities and state cleanup programs. Since the EPA's general policy is to implement Superfund and RCRA cleanups in a similar fashion, it is likely that the use of innovative technologies at RCRA sites is similar to use of innovative technologies in the Superfund program (EPA, 1993). A noncomprehensive review of remedies for contaminated ground water at 15 RCRA sites indicated that pump-and-treat systems were chosen at 14 of the sites. At the fifteenth site, access to contaminated ground water was restricted rather than requiring ground water treatment. At two of the sites, bioremediation systems were chosen to operate in conjunction with the pump-and-treat systems (Davis, 1995). In an audit of cleanups at federal facilities, the U.S. General Accounting Office found that "although EPA, Energy and Defense have spent substantial sums to develop waste cleanup technologies, few new technologies have found their way into cleanups" (Guerrero, 1995).

BARRIERS TO INNOVATION

Since the late 1980s, reports from a variety of organizations have indicated that significant barriers exist to development of remediation technologies for commercial markets. Early reports were produced by a federal advisory commission known as the Technology Innovation and Economics Committee, part of the National Advisory Council on Environmental Policy and Technology. This group, established in 1989, assessed the use of all types of environmental technologies, focusing primarily on pollution prevention and recycling technologies but also considering hazardous waste cleanup technologies. The group concluded that "current environmental statutes and federal regulations do not encourage the development of innovative technological solutions" (EPA, 1989). Similar conclusions emerged from events and reports of a variety of other organizations, including

- a series of workshops hosted by the EPA's Technology Innovation Office for environmental consultants, government regulators, and regulated industries (EPA, 1990, 1992);
- a workshop convened by a federal commission known as the Federal Advisory Committee to Develop On-Site Innovative Technologies (the "DOIT" committee) (Federal Advisory Committee to Develop On-Site Innovative Technologies, 1993); and
- a report issued by the National Commission on Superfund, established to develop a consensus among industries, environmental groups, and government officials about changes needed in the Superfund program (National Commission on Superfund, 1994).

Barriers to use of innovative technologies are complex and range from the inherent variability of the subsurface environment, to regulatory obstacles, conservatism on the part of hazardous waste site owners and their consultants, and lack of trustworthy data on technology performance. Much of this report focuses on developing credible data sets that can be used to compare innovative technologies against conventional ones and to transfer technology used at one site to another site without having to repeat all elements of the testing. However, as explained in Chapter 2, market and regulatory barriers must be addressed, as well, in order for a technology testing program to be effective. The technical problems associated with cleanup of contaminated ground water and soil at hazardous waste sites are far from solved, and there is a great deal of room for innovation, provided disincentives to innovate can be eliminated.

REFERENCES

ASTM (American Society for Testing and Materials). 1995. Standard Guide for Risk-Based Corrective Action Applied at Petroleum-Release Sites. E1739-95. West Conshohocken, Pa.: ASTM.

Begley, R. 1996. Risk-based remediation guidelines take hold. Environmental Science & Technology 30(10):438A-441A.

Chapelle, F. H. 1992. Ground-Water Microbiology and Geochemistry. New York: John Wiley & Sons.

Cohen, R. M., and J. W. Mercer. 1993. DNAPL Site Evaluation. Boca Raton, Fla.: C. K. Smoley.

CBO (Congressional Budget Office). 1995. Cleaning Up Defense Installations: Issues and Options. Washington, D.C.: Congressional Budget Office.

Davis, H. R. 1995. Unpublished data from the RCRA Corrective Action Program. Washington, D.C.: EPA, Office of Solid Waste.

Environmental Business International. 1995. Global Environmental Market and United States Industry Competitiveness Report. San Diego: Environmental Business International.

EPA. 1989. Report and Recommendations of the Technology Innovation and Economics Committee: Recommendations for Action on Technology Innovation. Washington, D.C.: EPA, Office of Cooperative Environmental Management.

EPA. 1990. Meeting Summary: Workshop on Developing an Action Agenda for the Use of Innovative Remedial Technologies by Consulting Engineers. Washington, D.C.: EPA, Technology Innovation Office.

EPA. 1992. Meeting Summary: Project Listen—Enhancing Technologies for Site Cleanup. Washington, D.C.: EPA, Technology Innovation Office.

EPA. 1993. Cleaning Up the Nation's Waste Sites: Markets and Technology Trends. EPA 542-R-92-012. Washington, D.C.: EPA, Office of Solid Waste and Emergency Response.

EPA. 1996. Innovative Treatment Technologies: Annual Status Report (Eighth Edition). EPA-542-R-96-010. Washington, D.C.: EPA, Office of Solid Waste and Emergency Response.

Federal Advisory Committee to Develop On-Site Innovative Technologies. 1993. Regulatory and Institutional Barriers Roundtable: Comment Draft Report from Regulatory Barriers Roundtable, San Francisco, October 20, 1993. Washington, D.C.: EPA, Technology Innovation Office.

Federal Facilities Policy Group. 1995. Improving Federal Facilities Cleanup. Washington, D.C.: Council on Environmental Quality, Office of Management and Budget.

GAO (General Accounting Office). 1995. Community Development: Reuse of Urban Industrial Sites. GAO/RCED-95-172. Washington, D.C.: GAO.

Guerrero, P. F. 1995. Federal Hazardous Waste Sites: Opportunities for More Cost-Effective Cleanups. GAO/T-RCED-95-188. Washington, D.C.: GAO.

GWRTAC. 1995. NETAC Selected to Operate National Ground-Water Remediation Technology Center (News Release). Pittsburgh, Pa.: Ground-Water Remediation Technologies Analysis Center (http://www.chmr.com/gwrtac).

Heath, R. C. 1980. Basic Elements of Ground-Water Hydrology with Reference to Conditions in North Carolina. U.S. Geological Survey Water Resources Investigations Open-File Report 80-44. Washington, D.C.: U.S. Government Printing Office.

Heath, R. C. 1983. Basic Ground-Water Hydrology. U.S. Geological Survey Water Supply Paper 2220. Washington, D.C.: U.S. Government Printing Office.

Laws, E. P. 1996. Letter to Superfund, RCRA, UST, and CEPP national policy managers, Federal Facilities Leadership Council, and brownfields coordinators regarding EPA initiatives to promote innovative technology in waste management programs. EPA, Washington, D.C. April 29. Letter.

MacDonald, J. A. 1994. Germ warfare. Garbage: The Independent Environmental Quarterly (Fall):52-57.

MacDonald, J. A., and M. C. Kavanaugh. 1994. Restoring contaminated groundwater: An achievable goal? Environmental Science & Technology 28(8):362A-368A.

MacDonald, J. A., and M. C. Kavanaugh. 1995. Superfund: The cleanup standard debate. Water Environment & Technology 7(2):55-61.

MacDonald, J. A., and B. E. Rittmann. 1993. Performance standards for in situ bioremediation. Environmental Science & Technology 27(10):1974-1979.

Mercer, J. W., D. C. Skipp, and D. Giffin. 1990. Basics of Pump-and-Treat Ground-Water Remediation Technology. EPA/600/8-90/003. Ada, Okla.: EPA.

National Commission on Superfund. 1994. Final Consensus Report. Keystone, Colo.: The Keystone Center.

NRC. 1993. In Situ Bioremediation: When Does It Work? Washington, D.C.: National Academy Press.

NRC. 1994. Alternatives for Ground Water Cleanup. Washington, D.C.: National Academy Press.

OTA (Office of Technology Assessment). 1989. Coming Clean: Superfund Problems Can Be Solved. PB90-142209. Springfield, Va.: National Technical Information Service.

OTA. 1995. State of the States on Brownfields: Programs for Cleanup and Reuse of Contaminated Sites. Springfield, Va.: National Technical Information Service.

Reichard, E., C. Cranor, R. Raucher, and G. Zapponi. 1990. Groundwater Contamination Risk Assessment: A Guide to Understanding and Managing Uncertainties. Washington, D.C.: International Association of Hydrological Sciences.

Russell, M., E. W. Colglazier, and M. R. English. 1991. Hazardous Waste Remediation: The Task Ahead. Knoxville: University of Tennessee, Waste Management Research and Education Institute.

Spitz, K., and J. Moreno. 1996. A Practical Guide to Groundwater and Solute Transport Modeling. New York: John Wiley & Sons.

Tremblay, D., D. Tulis, P. Kostecki, and K. Ewald. 1995. Innovation skyrockets at 50,000 LUST sites. Soil and Groundwater Cleanup 1995 (Dec.):6-13.

2

Market-Based Approaches for Stimulating Remediation Technology Development

The market for new environmental technologies, including those for contaminated site cleanup, peaked in 1990. During the 1980s, investors had flocked to the remediation technologies market in response to major new laws (the 1980 Comprehensive Environmental Response, Compensation, and Liability Act and the 1984 amendments to the Resource Conservation and Recovery Act) requiring cleanup of the nation's waste sites. Investors assumed that the very large number of contaminated sites, combined with strict federal enforcement of the new regulations, would create a large market for innovative cleanup technologies and saw the potential for high returns from environmental investments.

Investors' predictions about the remediation technologies market were not borne out. As shown in Figure 2-1, by 1993 the strength of stocks in environmental companies, including those involved in remediation, had plummeted to a less than half of its peak value, and it has continued to decline. Despite the billions of dollars per year spent on remediation and other environmental programs, companies have struggled to bring new remediation technologies to the market. The lack of affordable commercial technologies has, in turn, led to resistance to attempting to clean up sites and to a push for the use of risk-based corrective action approaches (see Chapter 1). This move to limit the number of site cleanups based on technical feasibility of cleanup and/or risk factors has further weakened the market for remediation technologies.

This chapter explains why the remediation technologies market has been so much weaker than initially predicted. It recommends ways to increase the market demand for innovative remediation technologies by moving to a system that relies on market pull, rather than regulatory push, to guide technology selection. Under this new system, the costs of leaving contaminants in place and delaying

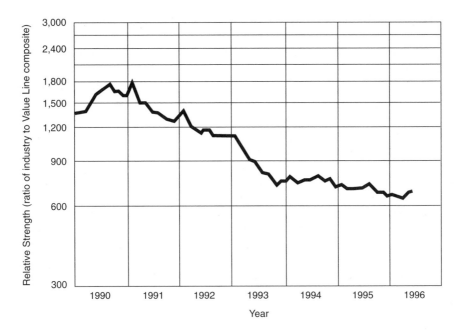

FIGURE 2-1   Relative strength of environmental company stocks traded on the U.S. stock exchange during 1990–1996. The ratio shown is stock value for the candidate industry normalized to the Value Line composite stock value, with August 1971 serving as the base index of 100. SOURCE: Reprinted, with permission, from Value Line, Inc. (1996). © 1996 by Value Line, Inc.

remediation would be explicit and consistent, allowing owners of contaminated sites to compare these costs with those of installing a remediation technology and cleaning up the site.

FATE OF INNOVATIVE TECHNOLOGY VENDORS

While successful examples of the introduction of innovative technologies for waste site remediation exist, such examples are relatively rare. As explained in Chapter 1, the range of technologies used to clean up hazardous waste sites is still quite limited. For example, although 14 general types of innovative technologies have been chosen for cleanup of contaminated soil at Superfund sites, 4 of these technologies—soil vapor extraction, thermal desorption, ex situ bioremediation, and in situ bioremediation—account for the bulk of projects. All other types of innovative soil cleanup remedies were selected for a total of fewer than 6 percent of Superfund sites where soil cleanup is under way (see Figure 1-7 in Chapter 1). Innovative remedies for ground water contamination are used at only 6 percent of

Superfund sites (EPA, 1996a). Thus, considering the magnitude of the waste site remediation problem, the number of types of innovative technologies in use is small, and vendors of innovative technologies have had difficulty capturing market share.

Start-up companies founded on trying to market new remediation technologies have generally fared poorly. Table 2-1 shows the recent stock value of the seven companies that have gone public based on marketing of a technology for waste site remediation. As shown in the table, the stock price of six of these seven companies has dropped since the initial public offering.

In today's market, remediation technologies are generally provided by large consulting firms offering a diverse range of environmental services, rather than by small companies offering "boutique" services focused on a specific niche of the market. There is little possibility that a firm offering a specialized technology will survive. The remediation industry is increasingly consolidating and diversifying, with fewer and fewer firms available to provide remediation technologies. In 1995, for example, there were 55 acquisitions of U.S. environmental services firms and in 1996 there were 73 acquisitions (ENR, 1996b). The one company in

TABLE 2-1 Stock Value of Selected Remediation Technology Companies

Company	Technology Area	Initial Public Offering			Recent Price per Share ($)
		Date	Price per Share ($)		
Envirogen	Biotreatment applications	8/92	7		$2^{3}/_{4}$
Molten Metal Technology	Catalytic extraction processes	2/93	16		15
Ensys Environmental Products	Immunoassay products	10/93	10		$1^{1}/_{2}$
Purus, Inc.	VOC control	11/93	14		$4^{3}/_{8}$
Thermo Remediation, Inc.	Thermal processing	12/93	8		10
Conversion Technologies International	Vitrification technology	5/96	$4^{1}/_{2}$		2
Thermatrix	Flameless thermal oxidation	6/96	$12^{1}/_{2}$		$9^{1}/_{4}$

NOTE: Initial public offering prices for the first five companies listed are quoted to the nearest point. Recent share prices are as of November 19, 1996.

Table 2-1 whose stock value increased after the initial public offering, Thermo Remediation, markets a system for thermally treating petroleum-contaminated soils but has increasingly diversified its services. It is affiliated with companies that collect and recycle used motor oil, provide wastewater processing services, and remove radioactive contaminants from soils. In addition, in December 1995 Thermo Remediation acquired Remediation Technologies Inc., an engineering/ construction firm that provides a range of environmental services.

While research continually generates new ideas for how to clean up contaminated sites, small firms that have been founded based on new research ideas have not fared well. The inability of small firms with new ideas to survive discourages innovation. Large, service-oriented firms generally provide their clients with "safe" technologies rather than risking a new approach that might perform better than the traditional one but that also has a chance of failing.

ELEMENTS OF THE REMEDIATION TECHNOLOGY MARKET

As is evident from the lack of success of new ventures in bringing remediation technologies to the market, the remediation market is difficult to enter. In part, this is a result of barriers to innovation that are a construct of the regulatory process, but it is also in part a result of the inherent fragmentation of the remediation market. The market is fragmented by client type and, more importantly, by site type.

The clients for remediation technologies can be grouped into two broad categories: (1) private sector, including a broad range of company types and sizes, and (2) public sector, including federal agencies, primarily the Departments of Defense (DOD) and Energy (DOE). About one-third of the remediation market consists of contaminated sites owned by the federal government (Russell et al., 1991), while the remainder consists of privately owned sites. Within the private-sector market, there is wide variation by client type and site size. For example, most of the contaminated sites shown in Table 1-2 in Chapter 1 are leaking underground storage tanks, many of them owned by small gasoline stations, while the larger, more complex sites are usually owned by large industries or groups of industries. Similarly, the characteristics of the public sector remediation market vary because of the wide variation in the agencies (ranging from the U.S. Department of Agriculture to the DOE) responsible for contaminated sites. The factors of importance to one public agency differ from those important to other agencies, which, in turn, differ from those important to large private corporations, which differ from those of greatest importance to gasoline stations or other small enterprises with contaminated sites. Further complicating matters, clients (whether public agencies or private industries) are usually represented by consultants, who may have their own concerns about technology performance. Thus, technology vendors need to develop different sales strategies, depending on the client and the client's consultant.

A much more difficult problem for remediation technology vendors is the fragmentation of the remediation market based on site type. A technology that works well for cleaning up a particular contaminant, such as petroleum hydrocarbons, in a particular geologic setting, such as a sandy aquifer, may not work at all for the same contaminant in a different geologic setting, such as a fractured rock aquifer. As a result, in the remediation business it is often not possible to market "widgets" that the client can simply plug in and use. Almost always, those "widgets" must be accompanied by significant technical expertise on how to apply the system in the site-specific setting. Such expertise is usually provided by consultants. Technology vendors therefore must either diversify to provide consulting services themselves or must convince consultants hired by their client that the new technology has merit.

REGULATORY BARRIERS TO INNOVATION

The regulatory structure for implementing hazardous waste cleanups, especially at Superfund and Resource Conservation and Recovery Act (RCRA) sites, has added to the inherent difficulties that remediation technology vendors face in bringing new products to the market. The fundamental problem with these programs is that they rely on regulatory push rather than market pull to create demand. The process of technology selection is strictly regulated. At the same time, the penalties for failing to initiate remediation promptly are insufficient. The result is that companies responsible for cleanups often delay remediation rather than trying new technologies because they perceive no economic gain from accelerated cleanup. Providers of new technologies have trouble staying in business while awaiting client and regulatory acceptance of their processes. Although the federal government has sponsored numerous initiatives, from the Superfund Innovative Technologies Evaluation program to the Strategic Environmental Research and Development Program, to promote innovative technology development (see Chapter 5), without the necessary market demand in place the technologies developed under these programs will not become widely used.

At Superfund and RCRA sites, technology selection is often a negotiated process between the regulators and the regulated. The market for new technologies becomes stifled for two reasons. First, regulatory restrictions limit a customer's freedom to choose a remediation technology and adapt the remedy over time as new technologies emerge. Second, the ability to arbitrate a cleanup often removes the incentive for improved solutions. In other market sectors, such as the computer and information technology industry, customers create demand for new technologies because they have freedom to choose and desire improved solutions. New technologies are then developed through a process of trial and error. New companies depend on early users to "de-bug" a new technology and use that information to make the necessary adjustments and improvements before the technology or product is released for commercialization. This type of gradual diffusion and adoption of technology has not worked effectively, except in a few cases, in increasing the market share for innovative remediation technologies. The Superfund and RCRA corrective action programs leave little room for customer (or consultant) choice and no room for a "try as you go" concept. Regulators must "sign off" on the customer's choice of a technology through an official Superfund record of decision or RCRA corrective action plan. Mechanisms for adjusting the remedy once it is officially approved are bureaucratically cumbersome and provide a disincentive to change the selected remedy even if a much better solution evolves.

In many instances, it is less costly for a company to delay remediation through litigation than to select a technology and begin cleanup. The incentive to delay rather than begin cleanup reduces market demand for remediation technologies. For example, the Congressional Budget Office has estimated that the average cost to clean up a private-sector Superfund site is $24.7 million (Congressional Budget Office, 1994b). Yet, analysis of corporate annual reports and financial statements shows that companies typically report a liability of about $1 million for sites where they have not yet begun cleanup. Thus, many companies are faced with a choice of cleaning up and taking an immediate cash drain of, on average, $25 million or carrying a $1 million annual liability with some litigation and assessment costs. Under these circumstances, there is little question that delay is the preferred alternative, because spending for full remediation might cause a company to lose a major portion, or all, of its cash reserves. The RAND Corporation sampled 108 firms with annual revenues of less than $20 billion involved in the cleanup of Superfund sites and found that transaction costs associated with legal work accounted for an average of 21 percent of these firms' spending at Superfund sites; spending on transaction costs exceeded 60 percent of the cost share for more than one-third of the firms (Dixon et al., 1993). The U.S. General Accounting Office (GAO) has estimated that Fortune 500 companies spend fully a third of their costs at Superfund sites on legal expenses such as disputing cost shares with other potentially responsible parties and negotiating remedy selection with the Environmental Protection Agency (EPA) (GAO, 1994b). At many sites, this extensive litigation serves the purpose of delaying remediation expenses.

Adding to the incentive to delay are the possibility that waste site cleanup regulations will change and the inconsistent enforcement of existing regulations. Like other laws and regulations implementing them, Superfund and RCRA have been subject to political swings. For example, current political pressure for Superfund reform is tending toward less stringent cleanup standards and requiring cleanup at a narrower range of sites. In 1995, a bill for Superfund reform, H.R. 2500, was introduced to the U.S. House of Representatives that would eliminate the requirement to consider all applicable or relevant and appropriate requirements in setting cleanup goals, meaning essentially that goals would be relaxed; eliminate the preference for treating contaminated water rather than developing alternative water supplies; make responsible parties liable only for damages to resources that are currently being used; and limit the number of new Superfund sites identified each year to 30. Other bills introduced to Congress have proposed eliminating or limiting retroactive liability for contaminated sites. Some organizations are lobbying to incorporate in Superfund the risk-based corrective action standards being developed by the American Society for Testing and Materials; use of these standards would substantially decrease the number of sites at which active cleanup would be required. Taken together, these proposals would significantly reduce the level of cleanup responsible parties are liable to undertake. If companies and responsible government agencies knew for certain that existing cleanup standards would be strictly enforced, there would be an incentive to clean up sites sooner. However, shifting political forces and changing legislative agendas, combined with a lack of sufficient penalties for failing to comply with existing regulations, reward those who wait for political relief.

Further encouraging delay in cleanup, economic incentives for carrying out remediation are lacking under current policies. Companies perceive remediation as a tax on earnings and a drain on their bottom line, rather than as an activity undertaken in the company's economic self interest. Although remediation ex-

TABLE 2-2 Earnings Used to Support Environmental Remediation at Selected U.S. Corporations (1994, millions of dollars)

Company	Sales	Earnings	Remediation Expenses	Percent of Earnings to Support Remediation
Allied Signal	12,817	759	66	8.7
Amoco	30,362	1,789	119	6.7
ARCO	17,199	919	160	17.4
Chevron	35,130	1,693	182	10.8
DuPont	39,333	2,727	91	3.3
General Electric	60,109	4,726	98	2.1
General Motors	154,951	4,900	105	2.1
Monsanto	8,272	622	52	8.4
Sun Company	9,818	90	60	66.7
TOTAL	367,991	18,225	933	5.1

SOURCE: Actual expenses as reported in 1994 corporate annual reports and 10-K statements.

penses have a significant impact on the profit margins of many large U.S. corporations, improvement in remediation technologies has not been linked to improved financial performance. Table 2-2 shows that for several large corporations, an average of about 5 percent of corporate earnings goes toward supporting remediation expenses. Yet, managers at companies often are unaware of the true cost of their remediation activities because they frequently do not account for remediation costs and report them to shareholders. For example, a Price Waterhouse survey of securities issuers in 1992 found that as many as 62 percent of responding companies had known environmental liabilities that they had not yet recorded in their financial statements (Blackwelder, 1996).

In the absence of assessing the liability for cleaning up contaminated sites and posting this liability on corporate balance sheets, there is no economic driver for improved remediation. As an analogy, companies are required to assess and fully report future pension and health care liabilities, providing an incentive to control pension and health care costs. This incentive is lacking for remediation. To the contrary, if a company were to voluntarily assess all of its future remediation costs and post the total on its balance sheet, the value of the company would be reduced, creating a disadvantage relative to companies that do not report this liability. Companies cannot show that spending more resources on remediation will result in improved earnings or reduced liabilities. It therefore becomes difficult for companies and their consultants or advisors to see the financial benefit of early remediation. For investors, the lack of financial drivers is especially troublesome because capital providers are primarily in the business of creating the highest possible rate of return for a given level of risk. Without being able to identify the value provided to the customer by a new technology, investors have difficulty estimating their potential investment returns and tend to shy away from the remediation sector to more familiar markets. As shown in Table 2-3, venture capital investment in environmental technologies (including remediation and other environmental technologies) is more than an order of magnitude lower than investment in other modern technologies such as biotechnologies and communications systems. While total venture capital investments have nearly doubled since 1992, venture capital investment in environmental technologies has declined by nearly 70 percent.

In part because of the incentives to delay remediation and in part because of the long series of regulatory steps involved in selecting a cleanup remedy for a site, the time line for selecting and installing a remediation technology can be very long and can vary unpredictably from site to site. According to the Congressional Budget Office, for example, the average time between when a site is proposed for listing on the Superfund National Priorities List (NPL) and completion of construction of the cleanup remedy was 12 years for the first 1,249 sites on the NPL (Congressional Budget Office, 1994a). Although the EPA in the early 1990s instituted administrative reforms to try to speed cleanup of NPL sites, a recent GAO analysis showed that cleanup completion times increased between 1989

TABLE 2-3   Venture Capital Disbursements, 1991–1995

Industry	Amount Invested ($ millions)				
	1991	1992	1993	1994	1995
Communications and networking	608.7	588.1	881.4	875.5	1,375.7
Electronics and computers	370.9	384.3	274.0	474.8	463.8
Semiconductors and components	163.4	189.9	244.5	189.3	301.7
Software and information services	461.9	547.2	528.0	745.7	1,239.1
Medical compounds	498.3	700.4	715.0	720.5	715.5
Medical devices and equipment	474.5	468.8	421.9	463.3	607.1
Health care services	141.9	221.0	322.6	326.2	492.6
Retailing and consumer products	267.8	213.5	544.5	529.1	1,206.9
Environmental	64.6	93.8	65.8	54.5	29.0
Other	276.1	574.9	502.1	641.4	1,000.1
TOTAL	3,328.1	3,981.9	4,499.8	5,020.3	7,431.5

SOURCE: VentureOne Corp., 1996.

and 1996 (Guerrero, 1997). Figure 2-2 shows the remediation time line for a site that provides an extreme example of delay in remediation: a RCRA site where, more than 20 years after contamination was discovered, a final remedy is not yet in place.

While remediation of all but the simplest sites requires a significant investment of time because of the technical difficulty of site characterization and remediation technology design, the financial disincentives to initiate remediation and time-consuming bureaucratic procedures can greatly increase the uncertainties associated with predicting the timing of remediation projects, as shown in Figure 2-2. Unpredictable time delays make it very difficult for technology developers and funders to forecast cash flow. Start-up technology providers have gone out of business for lack of cash flow while waiting for final regulatory approval to use their technology at a large enough number of sites to stay solvent (see Box 2-1).

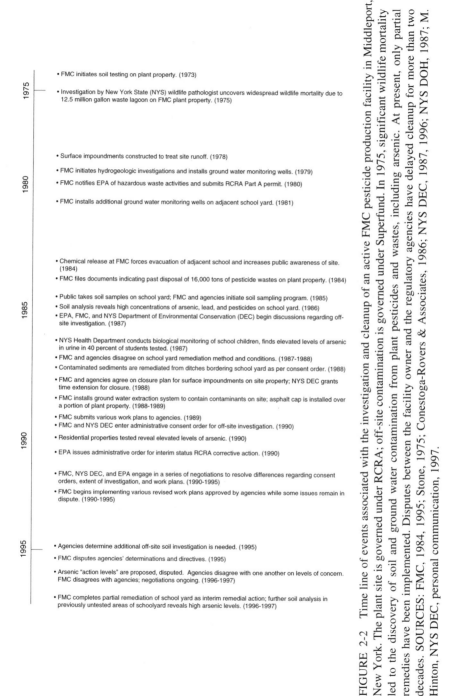

FIGURE 2-2 Time line of events associated with the investigation and cleanup of an active FMC pesticide production facility in Middleport, New York. The plant site is governed under RCRA; off-site contamination is governed under Superfund. In 1975, significant wildlife mortality led to the discovery of soil and ground water contamination from plant pesticides and wastes, including arsenic. At present, only partial remedies have been implemented. Disputes between the facility owner and the regulatory agencies have delayed cleanup for more than two decades. SOURCES: FMC, 1984, 1995; Stone, 1975; Conestoga-Rovers & Associates, 1986; NYS DEC, 1987, 1996; NYS DOH, 1987; M. Hinton, NYS DEC, personal communication, 1997.

BOX 2-1
GRC Environmental: Cash Flow Problems Due to Slow
Acceptance of an Innovative Remediation Technology

In the late 1980s, a company known as GRC Environmental had developed a patented technology for cleaning up polychlorinated biphenyls (PCBs) and dioxin in soil. The technology and company appeared destined for success. GRC had demonstrated its technology at the pilot scale. PCB and dioxin cleanup was scheduled to occur at a large number of sites. The technology's main competitor, incineration, was losing favor with regulators and the public. The technology, an alkaline substitution process that operates at elevated temperature, had a particular advantage over incineration in that it produced no harmful byproducts. However, although the company was able to obtain regulatory approval to use its technology at one site, it went out of business for lack of approval to use the method at a large enough number of sites to maintain a consistent cash flow (Houlihan, 1995).

GRC experienced great difficulty in landing its first job contract. The venture capitalists who funded the company attributed this difficulty to distrust that the technology would be approved by regulators. GRC finally secured a contract for a first job, a Superfund site in Houston, only because the site presented so many challenges that the primary contractors had been unsuccessful in achieving cleanup goals (Houlihan, 1995). One of the primary technical difficulties at the site was that the soil was clay, which is very difficult to clean. Further, during the cleanup, Houston experienced its heaviest rain in years, interfering with equipment operation.

Technically, GRC's cleanup of the Houston site was a success in that it restored the soil to the satisfaction of regulators. However, the job was a financial failure (Houlihan, 1995). Because of the lack of additional contracts to provide cash flow and the backlog of expenses from the Houston site, GRC had to file for bankruptcy. GRC had invested too much of its capital to construct the system in Houston, and it had been unable to obtain the additional jobs that could have kept the company solvent even with losses at the Houston site.

The lack of predictable timing is of particular concern to investors because they are unable to project investment returns. Worse, there is a disincentive for investors to provide funds early in the technology development cycle because the technology does not appreciate in value until just before it becomes commonly accepted. Thus, there is insufficient reward for the additional risk of having provided capital at the early stage of development.

Like the remedy selection time line, the end point that a technology must achieve and the pathway for achieving regulatory approval are unclear. While drinking water standards historically have been selected as cleanup goals at most sites (National Research Council, 1994), this is not always the case. The GAO reviewed cleanup standards for ground water and soil in 21 states in 1996 and found that the standards vary widely (GAO, 1996b). In an earlier review, the GAO found that soil cleanup goals for polycyclic aromatic hydrocarbons at 14 Superfund sites ranged from 0.19 to 700 parts per million (Hembra, 1992). The variation was a function of what decision regulators made about the future use of the site, but the factors weighed in making this decision were unclear. For example, cleanup standards varied among sites that were equally near to residential areas.

Current Superfund and RCRA regulations allow for wide discretion by regulators about what level of cleanup should be required at a given site, and individual regulators may reach their decisions about cleanup end points using different methods. Administration of Superfund and RCRA is carried out by EPA's ten regional offices, and each office has somewhat different strategies for setting cleanup goals. Furthermore, sites having the same geophysical and contaminant characteristics may be subject to different legal requirements depending on whether the site is a federal Superfund site, included on a state list of contaminated sites, or governed by the RCRA corrective action program. For example, in a review of compliance with Superfund and RCRA requirements at DOE facilities, the GAO concluded that although the Superfund and RCRA programs have broadly similar objectives, "the two programs differ in their highly detailed sets of procedural regulations and guidelines and in the particulars of their implementation" (GAO, 1994a). According to the GAO, for example, remedies for RCRA sites are typically selected for relatively small unit areas, while Superfund cleanups are generally regulated over broad geographic areas. Without clear, consistent regulatory requirements on how to receive approval for a remediation technology, it is difficult for technology developers to prove to potential customers that their technology is acceptable to regulators, even if the developer has cost and performance data. Thus, technology developers and customers may find themselves in a Catch 22: the customer wants to be assured that the technology will be accepted by regulators, but the regulator wants to see the technology in operation before providing the permit, meaning that the developer first needs to sell the technology to a customer.

Lack of consistent performance standards for remediation technologies, combined with inherent uncertainties about technology performance, has resulted in customers most often seeking a technology that can achieve regulatory compliance, rather than one that can reach a specific end point. Remediation technologies are thus more of a legal product than a technological one, and there is little or no premium for improved solutions to subsurface contamination problems. Either a technology meets the approval of regulators and has high value, or it does not meet regulatory approval and has no value. That is, a company must meet regulatory approval for a given site, and anything more does not provide additional value to the site owner. If a site owner has received regulatory approval for a remediation strategy at a site and new technologies become available that can improve cleanup at equal or lower costs, there is no reason for the site owner to engage in additional remediation once legal compliance is achieved.

In summary, the processes for implementing Superfund and RCRA have reduced the potential market demand for new remediation technologies. Under these programs, delay is preferable to initiating cleanup, in part because companies perceive that remediation is not in their self interest. The time line for selecting remediation technologies is lengthy and unpredictable, making it hard for technology start-up companies to stay in business while they await approval of their first customer's contract. The end points that a remediation technology must achieve are often negotiated and vary from site to site, and current regulatory policy provides no premium for solutions that exceed the regulatory requirements. Customer freedom to choose new technologies and to adapt solutions over time as better technologies emerge is highly restricted. All of these factors contribute to the weakening of the market for waste site remediation technologies.

It is important to note that the market for new technologies is stronger in the underground storage tank (UST) cleanup program than in the Superfund and RCRA corrective action programs. In part, the relative strength of the UST market reflects the fact that spills from underground storage tanks are much simpler to clean up than contamination at RCRA and Superfund sites (National Research Council, 1994). Underground storage tanks typically contain just one type of contaminant (usually petroleum hydrocarbons, which are relatively easy to clean up), and they affect a relatively small area. However, the greater use of innovative technologies in the UST market also reflects the greater freedom to choose innovative technologies allowed in the regulations governing these cleanups. UST cleanups are controlled at the state and local levels, rather than the federal level. Customers have freedom to choose a remediation technology for UST sites, and the job of regulators is simply to ensure that the site has been cleaned up to the required level. State regulators have expressed concern that inadequate oversight of UST cleanups due to the rush to remediate and redevelop these sites has in some cases resulted in incomplete cleanup of the sites. To prevent such situations, customer freedom to choose remediation technologies must be accompanied by strong regulatory enforcement of cleanup goals.

OTHER BARRIERS TO INNOVATION

Not just regulatory programs but also the actions of remediation clients have frustrated attempts to commercialize innovative remediation technologies.

In the private-sector remediation market, companies can be hesitant to share information about their contaminated sites. This lack of information sharing makes it very difficult for technology vendors to predict the potential size of the market for their product and to establish sites to which future clients can be referred for evidence of the technology's performance. Very few companies are completely open about their remediation needs, given the negative public image and increased regulatory scrutiny this would create. Thus, technology providers and investors have poor information about the distributions of sites having characteristics suitable for application of their technologies. The EPA's Technology Innovation Office has prepared reports that assess market opportunities for innovative remediation technologies in the southeastern and mid-Atlantic states (EPA, 1995b, 1996b), but these reports, while a useful starting point, lack the detailed information about contaminants, type and volume of contaminated media, and hydrogeologic characteristics of the sites that are essential for allowing technology developers to easily identify whether their technologies might be applicable at a given site. Lack of information about the size of the various market segments makes it very difficult for technology developers and investors to predict their potential sales. In addition, because many companies are concerned about the negative public image associated with having contaminated property, few are willing to have their property used as a reference site for a remediation technology. For new technologies, establishing a "blue chip" list of customers is critical. The presence of such customers can attest to the value of the new technology, and the established sites where the technology has been used can serve as reference sites for other customers. It is especially difficult for remediation companies to establish a list of reference sites in the private-sector market.

In the public-sector market, inadequate cost containment has decreased the incentives for selecting innovative technologies. Often, federal remediation contractors are placed on "auto pilot" after being awarded the cleanup contract on a cost-reimbursable basis, so there is little incentive for cost effectiveness (GAO, 1995b). According to an audit by the GAO (1995b), cost overruns are common to remediation efforts at federal sites, due in part to inadequate oversight of contractors. GAO found evidence of fraud, waste, and abuse by federal remediation contractors (GAO, 1995b). With no incentive to reduce costs, there is no incentive to search for new solutions.

In summary, lack of information sharing in the private-sector remediation market and inadequate control over costs in the public-sector remediation market create barriers to innovation that add to those that are inherent to the market itself and those created by the regulatory process.

CHARACTERISTICS OF
FLOURISHING TECHNOLOGY MARKETS

The amount of venture capital invested in a given market is an indication of the perceived health of the market and the drive for innovation. Venture capital investors to a great extent seek the path of least resistance. They try to achieve the greatest return possible for any given level of risk. Venture capital thus flows quickly from one industry to another, depending on which industries are perceived as offering the greatest potential for profits from new technologies. In considering how to reinvigorate the market for innovative solutions to hazardous waste site remediation, it is useful first to outline the characteristics of technology sectors that attract relatively large amounts of venture capital (see, for example, Table 2-3).

In general, the industries most successful in attracting venture capital, such as the software and medical drug markets, have the following characteristics:

• *Market is driven by performance and cost:* Above all else, investors seek markets and industries where new technology can be leveraged to create a product or service that generates measurable value to the customer. That is, the customer must receive returns that they perceive as greater than the cost of the product. The enormous growth in the computer and software industry over the last two decades is a clear example of this phenomenon. Investors continue to pour capital into new computer companies to develop new products to solve the same problem either less expensively or more efficiently, or to enable the customer to perform a task that was not possible before. The market rewards those who are able to provide the product or service most efficiently. The trade-off between cost and performance is clear: improved performance costs more, and there is a large market for slightly lower performance at discounted prices.

• *Customers have freedom to choose:* The most efficient markets are characterized by customer freedom to choose. In order to survive in free markets, companies must continue to develop improved products because the consumer has the freedom to choose any product or service. The consumer can at any time abandon one product in favor of a more effective or equally effective but less expensive alternative.

• *Rewards justify the risk of innovation:* Technology innovation is traditionally encouraged by the opportunity to generate financial returns commensurate with the risk level of the venture. Greater risk generally carries with it the opportunity to create greater rewards. The typical remediation start-up company is not competing for investment capital against a number of similar remediation companies but against all other start-up companies; the risk profile of remediation firms needs to be less than or equal to the risk profile of other types of companies in which investors might choose to place their capital.

• *Market is quantifiable:* In successful markets, extensive information about the size and characteristics of the market is available, enabling entrepreneurs and investors to predict their potential returns and tailor their technologies to meet market needs. Similarly, extensive information about performance and cost of existing technologies is available, allowing entrepreneurs and investors to gauge market needs.

• *Time to market is short and predictable:* Successful markets tend to have a predictable and understandable path to the customer. The "rules" are known,

---

BOX 2-2
The Pharmaceutical Industry: How a High Level of
Regulation Can Coexist with Innovation

To obtain regulatory approval and enter the marketplace, a company with a new pharmaceutical drug must submit a "new drug application" (NDA) to the Food and Drug Administration (FDA). The NDA specifies manufacturing practices and contains safety and efficacy data based on a prescribed set of preclinical tests and three phases of clinical trials. Failure at any one of these stages can derail the project, and each new success brings new sources of funding. Before the FDA approves the NDA, it usually requires more work to better justify efficacy indications or potential side effects. The indications and side effects are then communicated to the market on labels and other information accompanying the approved drug. Any change in manufacturing practices requires prior approval of the FDA.

While the level of government scrutiny in the drug industry is substantial, the hurdles that a new product must clear prior to regulatory approval are consistent and well established. Further, the approval process provides assurance to buyers and end users of drugs that the products are safe and effective, and this assurance helps underpin the market.

Most pharmaceutical companies invest about 10 to 20 percent of their sales in R&D. The potential reward for this enormous R&D investment is the infrequently discovered "blockbuster" drug that may bring in excess of a billion dollars per year in sales over the remaining patent life.

---

and if the market is regulated, the regulations are known and do not change significantly or rapidly over time. For example, the pharmaceutical industry is one of the most regulated markets, and one would expect it to be less attractive to investors because of the government approvals needed to bring a product to market. However, this sector is very well financed, despite the regulations, in part because the steps needed to obtain regulatory approval are clear (see Box 2-2). In rare cases, innovative remediation technology ventures have succeeded in commercializing their technologies by obtaining initial sales in markets unrelated to remediation in which the regulatory expectations are clear; Box 2-3 describes an example of one such venture, Thermatrix.

   • *Level of competition is high, and change is rapid:* Constant change in performance and cost attracts capital and investors in spite of the relatively high number of companies that do not succeed. This is because over time, investors can earn sufficient returns on their investments on a portfolio basis, in which multiple investments are made in the same category and individual losses are covered by successes. This portfolio approach is critical to overall return. The

---

BOX 2-3
Thermatrix, Inc.: Market Entry Through a Sector with
Clear Regulatory Guidelines

Thermatrix, Inc., owns rights to a technology that uses high heat in a porous ceramic medium to destroy contaminants and is a substitute for incineration (Jarosch et al., 1995; Schofield, 1995). The technology was originally developed for petroleum extraction in the oil shale program at Lawrence Livermore National Laboratory. After spending $25 million to develop the technology, the DOE abandoned it when the oil crisis ended and petroleum prices stabilized (Schofield, 1995). Three Livermore scientists left the lab to commercialize their invention. By the end of 1991, however, their company was insolvent and owed $2.5 million to creditors (Schofield, 1995).

New management took over in 1992. Although the most promising application for the Thermatrix technology was site remediation, entering this market was seen as too risky because (1) the time to market was too long given the company's debts and minimal available capital; (2) no funding was available for demonstration projects; and (3) consultants, rather than clients, usually choose the remediation technology and have a preference for tested approaches because of their lower risk (Schofield, 1995). Therefore, company managers targeted the air pollution market first. The company obtained several customer orders based on regulatory receptivity to a viable alternative to incineration and corporate frustration with the permitting process for incinerators. The company provided money-back guarantees to clients to avoid having dissatisfied customers who could show evidence of a Thermatrix system that was not performing up to specifications.

The next step was to enter the site remediation business. The first client was General Electric (GE), which is responsible for a Superfund site contaminated with PCBs where the record of decision specifies incineration, at a bid cost of $77 million (Schofield, 1995). Thermatrix approached GE managers directly and explained that the company's technology could destroy PCBs on the GE site for $6 million. A contract was entered to test the technology. Other applications followed. Capital for growth became more readily available. The company went public in June 1996 (see Table 2-1).

---

sectors of the economy that are the most competitive are the ones in which change and technical innovation and development occur most rapidly. For example, the U.S. semiconductor industry is the world's largest supplier of semiconductors, but capital equipment in this industry has a shorter life span than in any other sector of the economy. New chip manufacturing methods force yesterday's technology into obsolescence on average every 18 months. For example, Intel introduced its first pentium chip in 1993 and by 1996 was working on the third generation of this chip. Despite this short life cycle and the high cost (on the order of $1 billion) of capital equipment needed to build a new manufacturing line, this industry has no problems attracting investment capital because of the large investment returns it offers. Thus, it is not change, but uncertainty and lack of success, that repels investor capital.

• *Prior ventures have proven successful:* Above all else, what attracts capital (and people) to markets are successes. If a sector of the economy generates significant investment returns, capital will flow to that sector. Investors, like people in general, tend to follow others who have achieved success. Successes attract new capital and people, which in turn creates new companies, which in turn creates new successes. This phenomenon is responsible for the growth in high technology start-up companies concentrated in Silicon Valley, California, and along Boston's Route 128.

• *Business models are well developed:* Investors, from venture capitalists to corporate investors and public underwriters, seek to understand the business plan that a new company will follow to exploit the value of its technology. The model is the critical blueprint that allows the entrepreneur to pull together technology and capital to build a company. For example, the success of Regenesis Bioremediation Products, Inc., in developing its oxygen release compound (see Box 2-4) was in part due to a carefully developed business plan that the company prepared prior to offering the product to the market. Such models are rare in the remediation technology industry. Lack of clear business models is an indication of insufficient success in the marketplace

• *Experienced people are available to start new ventures:* Established and growing industries have a steady stream of people who have started new ventures before and want to do so again. The computer industry provides examples of entrepreneurs who establish new companies, build them to the level at which their expertise is no longer relevant, and then leave to start up a new venture. The existence of such entrepreneurs is one of the critical assets the financial community evaluates in deciding whether to fund a new venture.

Strong, innovative markets with the above characteristics attract public and private investment to support basic research, efforts at commercialization, and growth. The most critical stage for investment capital is that between the research and development (R&D) phase of technology development and successful commercialization. Investors typically refer to this phase as the "Valley of Death" (see Figure 2-3) because so many start-up firms fail in this transition. Significant capital is available for the early stage of technology development and R&D. For example, the DOE alone provided $47 million in research grants in 1996 through its Environmental Management Science Program (Renner, 1996) for basic research on subsurface contamination and remediation technologies. Indeed, the

---

BOX 2-4
Regenesis Bioremediation Products:
Carefully Developed Business Plan

The Oxygen Release Compound (ORC) of Regenesis Bioremediation Products, Inc., is one of the few remediation technologies that has been successfully commercialized in recent years. The success of Regenesis in commercializing this product was due in part to a carefully conceived business plan that the company developed prior to marketing its product and in part to backing from a major, financially solvent corporation during the critical stages of development.

ORC was developed to clean up low levels of dissolved petroleum hydrocarbons in the subsurface. The compound is a proprietary formulation of a metal peroxide that releases oxygen slowly when placed in the ground water environment. The oxygen thus released enhances aerobic bioremediation. The product is applied to the contaminated site in retrievable filter socks.

ORC (under another name) was originally developed for use in preventing gardening soil from becoming anaerobic, and it is sold in many home garden centers. The parent company that spawned ORC is a wealthy plant and garden company. The financial support of this solvent corporation ensured that the developers of ORC would have adequate resources during the critical stage between R&D and commercialization.

Development of ORC for the remediation market and incorporation of Regenesis was preceded by three years of product testing and demonstration. A scientific advisory board guided the research. A series of field trials verified product performance (Bianchi-Mosquera et al., 1994).

Prior to offering ORC to the market, the product developers contracted for a market study by Arthur D. Little. The study was completed in the summer of 1993, and outside capital was raised in the fall of 1994. Regenesis Bioremediation Products was incorporated in March 1994 to continue product development and commercialization.

Regenesis prepared a clearly focused market entry strategy based on the following principles:

• Sell ORC to environmental engineering and consulting firms, and avoid duplicating the types of service they provide.
• Focus on ground water applications.
• Establish the product in the United States before selling it abroad.
• Support firms interested in using ORC by helping them evaluate a site for ORC application.
• Assist consulting firms in selling ORC to the end user and in gaining regulatory approval.

Regenesis offered ORC to the market in February 1995. A year later, the product was being used in the remediation of 700 sites in the United States and Canada.

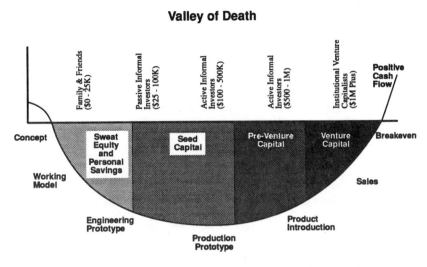

FIGURE 2-3 The "Valley of Death:" the stage between development of a new concept and successful commercialization of the concept. SOURCE: Adapted from SBA, 1994.

number of patents issued for technologies relating to "remediation" or "hazardous waste destruction or containment" increased from nearly zero in 1980 to more than 430 in 1993-1994, showing extensive innovation and suggesting a healthy level of R&D (see Table 2-4). Once a technology reaches a critical level of market acceptance, sufficient capital is usually available for growth and traditional commercial activity. In contrast, the stage between R&D and commercial sales is capital constrained because the amount of capital required to shift a technology from the lab bench to full-scale application is high, but the level of risk that it will fail is also high, making investors wary. Providers of capital at this critical "Valley of Death" stage are primarily driven by financial returns and not technical objectives; their sole purpose is to create wealth from the technology. It is often difficult for developers involved in the R&D phase, who are motivated by the desire to solve problems and prove the technical merit of their creation, to understand how to "sell" the technology to investors. Backing from large, financially solvent corporations during the critical "Valley of Death" stage was essential to the success of the innovative remediation ventures Regenesis, described in Box 2-4, and Geosafe, described in Box 2-5, in commercializing their products.

Once investors can see that they have a chance to recoup their investments, capital will return to the remediation marketplace. The question is how to adapt current regulatory policies and client perceptions to remove some of the barriers that have driven investment capital away from the remediation market. The regulations governing contaminated site cleanup cannot be eliminated. Without regulatory pressure, it is unlikely that most companies would pursue remediation on

TABLE 2-4   Remediation Technology Patents

Time Period	Patents Referring to "Remediation"	Patents in Class 588: Hazardous Waste Destruction or Containment
1976-1980	1	0
1981-1985	0	0
1986-1990	9	1
1991-1992	25	13
1993-1994	82	348
1995-1996 (8 months)	88	263

NOTE: Patent issuance data reflect patent applications filed an average of three years prior to the date of issuance. Thus, the trend toward increased patent activity in this sector began at the end of the 1980s, and a jump occurred in about 1990. The data for remediation are based on a search of patent titles and abstracts in the U.S. Patent and Trademark Office on-line data base. Class 588 is a recent addition to the classification system, so there may be prior relevant patents that were not reclassified into class 588.

their own. The key is to implement remediation laws in a way that provides economic incentives for companies and the federal government to implement cleanups as quickly as possible, rather than delaying remediation until some future time.

CREATING MARKET INCENTIVES FOR REMEDIATION TECHNOLOGY DEVELOPMENT

The principal change necessary to move to a market-oriented approach to remediation technology development is to take advantage of the power of financial self interest rather than relying on the force of regulation alone. The objective is to develop a market that is quantifiable, with reasonably well-defined risks and a commensurate opportunity to create financial returns from solving problems. That is, both the vendor *and* the customer must perceive financial benefit from improved remediation of contaminated properties, while still protecting the interest of the affected public in ensuring that sites are cleaned up. Capital will flow to the remediation technology market when it becomes evident that new technologies can create real value for customers.

The use of market forces in creating demand for remediation is the premise behind brownfield programs, in which state and local agencies provide incentives for redevelopment of contaminated urban industrial sites. For example, some states have recently developed provisions whereby they will forego lawsuits against new owners of contaminated property in economically distressed areas provided the owners follow the procedures of the state's voluntary cleanup program (GAO, 1995a). Brownfield programs create a high economic incentive to clean up these properties and redevelop them because many of the properties

A former fuel storage and handling facility in California was demolished to make way for a marina. As part of the process, 13,000 m³ of petroleum-contaminated soils were treated in above-ground bioremediation cells. Courtesy of Fluor Daniel GTI.

occupy extremely valuable parcels of urban land. Anecdotal evidence suggests that the market incentives driving the redevelopment of brownfields can lead to careful consideration of innovative remediation technologies, because property developers have a strong incentive to complete the remediation as quickly and effectively as possible so they can sell or lease the rehabilitated land. For example, at a brownfield site in downtown Wichita, Kansas, the city chose an innovative in situ treatment approach over a conventional pump-and-treat system in part because the innovative method could be more easily modified to improve performance (see Box 2-6).

While brownfield programs are increasingly being implemented at the local and state levels, economic incentives for remediation are still lacking in major sectors of the remediation market, particularly at sites regulated under Superfund and RCRA. In fact, lenders, environmental attorneys, and local officials have reported that fear of liability stemming from the Superfund program has discouraged brownfield redevelopment (GAO, 1996a). For example, in Allegheny County, Pennsylvania, some former steel mill sites are still idle in part because their owners prefer to keep the sites rather than risking that environmental assessments prior to sale will reveal contamination that they are liable for cleaning up

---

BOX 2-5
Geosafe: Financial Backing from Battelle
Essential for Survival

Geosafe Corporation is commercializing an in situ vitrification (ISV) technology initially developed by Battelle Pacific Northwest Laboratory for the DOE. The time from the formation of Geosafe until the company received its first commercial sale (5 years) was relatively long, and financial backing from Battelle has been essential for ensuring the company's survival during several downturns in the transition from R&D to commercialization (J. E. Hansen, Geosafe Corporation, personal communication, 1995).

ISV technology involves the in-place electric melting of earthen materials (EPA, 1995a). The earthen media itself serves as the containment for the melt. The melt occurs at a temperature range of 1600 to 2000°C. The high temperature causes the pyrolytic destruction and vapor-phase removal of organic contaminants. In addition, most metals are immobilized as oxides and incorporated into the vitrified product upon cooling.

ISV was conceptualized in 1980, and the initial patent for the process was filed in 1981 and issued in 1983. Battelle built a prototype for DOE in 1985 and demonstrated it in an application in 1987. Geosafe Corporation formed in 1988 to attempt to commercialize the process, while the DOE continued research focused on developing new applications within the DOE community. Geosafe demonstrated ISV in 1989 and made its first commercial sale during 1993. Since then, Geosafe has sold the technology for use at three additional sites in the United States. The company is also marketing the technology in Australia and Japan through ISV Japan Ltd. and Geosafe Australia (C. Timmerman, Geosafe Corporation, personal communication, 1997).

Geosafe encountered numerous difficulties prior to landing its first sale, and but for the financial backing of Battelle might not have survived. The company lost an initial project due to a competitor's claims against the technology, which were later disproved. Geosafe spent significant resources refuting those claims and repairing the damage they caused. In addition, Geosafe attempted to expand the market for ISV too early, before the fundamental processes underlying ISV were fully understood. This premature market entry led to a series of unexpected testing problems that consumed time and resources.

(GAO, 1995a). Banks have chosen not to foreclose on properties for fear of being held liable for remediation (GAO, 1995a, 1996a).

To shift the remediation technology market from one that is driven almost solely by regulations to one that captures the power of economic self interest, federal and state regulatory agencies need to pursue five types of initiatives. First,

---

BOX 2-6
Wichita Innovative Remediation Plan

In 1990, the Kansas Department of Health and Environment (KDHE) discovered chlorinated solvents in the aquifer directly under Wichita's downtown area (ENR, 1996a). Downtown commercial real estate activity came to a halt, and the city's downtown redevelopment program appeared doomed by the prospect of a Superfund listing. The site was immense, encompassing a 6.4-km (4-mile) long, 2.4-km (1.5-mile) wide area. Some $86 million worth of commercial and residential properties was affected in 8,000 parcels of land. Financial institutions, wary of Superfund liability, discontinued loans in the affected area, and one of Wichita's most important tax bases began to erode.

City officials decided to take responsibility for site cleanup rather than letting it proceed under the Superfund program, in effect becoming a remediation broker for the property owners. The city took a series of actions to increase the market value of the property by reducing concerns of potential buyers that they might be held liable for contamination. The overall program has been a financial success for the city. By agreeing to take on responsibility for remediation, the city has increased the revenue generated in the downtown area. The Old Town area is being revitalized with more than 20 new businesses. In this case, remediation was profitable for the city because of the high commercial value of the property.

In 1994, KDHE approved an innovative remedial plan (Olsen, 1996). The plan specified bioremediation as a possible remediation technology and required a pilot demonstration project. A "bio-curtain" comprising an in situ bioremediation trench and zero valent iron wall was tested. Bids were received for $18 million, $2 million more than conventional pump-and-treat technology. Although the innovative remedy was more expensive than the conventional one, the city selected it because of the potential that it could do a better job cleaning up the site than the conventional remedy and because of the potential for long-term cost savings to the city. The innovative approach, city engineers believed, could be more easily adapted to improve performance by optimizing microbial degradation of the contaminants. Also, they believed that costs for this approach would decrease after the initial application and that the city could implement the same strategy at other sites for a much lower cost.

---

economic incentives for remediation need to be created. Second, enforcement of regulations needs to be more consistent. Third, the regulatory process for selecting cleanup goals and remediation technologies needs to be more predictable. Fourth, complete information about the size and nature of all sectors of the remediation market, public and private, must be made available. Fifth, more opportu-

Sampling from monitoring wells at a field demonstration of an in situ bioremediation system for the treatment of chlorinated solvents in Wichita, Kansas (see Box 2-6). Courtesy of Roger Olsen, Camp Dresser & McKee.

nities need to be created to test innovative remediation technologies and verify their performance.

Economic Drivers

As described in this chapter, the remediation market is unique in its lack of economic drivers to accelerate the use of innovative technologies. If customers derived financial value and economic differentiation from improved remediation and accelerated cleanup, they would perceive remediation as an activity worth pursuing in part for their own self interest.

One way to create this self interest in large corporations is to improve corporate reporting of remediation liabilities to the Securities and Exchange Commission (SEC). The SEC currently has regulations for disclosure of environmental liabilities in documents such as quarterly and annual reports and security registration statements, but the guidelines are vague and subject to widely varying interpretation by accountants and corporate managers (see Box 2-7). As a result, companies often do not report remediation liabilities or fail to disclose the full magnitude of liabilities. Detailed guidelines should be developed for reporting of remediation liabilities, and sanctions for inaccurate or incomplete reporting should be increased. Consistent reporting of remediation liabilities to the SEC would help shift remediation from a cost center to a value-added activity. It would pro-

---

BOX 2-7
SEC Requirements for Reporting of
Environmental Liabilities

The SEC has general guidelines requiring publicly traded companies to disclose their liability for environmental remediation in documents such as quarterly and annual reports and security registration statements (SEC, 1993). However, these guidelines are subject to widely varying interpretation by accountants and corporate managers. Pressure by the SEC to disclose more has had limited effectiveness (Robb, 1993).

Specifically, SEC Regulation S-K, 17 C.F.R 229.10, contains the following provisions about environmental reporting (Roberts and Hohl, 1994; Roberts, 1994):

- Item 101 requires companies to include with the general description of their business any environmental expenditures that have a "material" effect on the business.
- Item 103 requires disclosure of pending or contemplated administrative or judicial environmental proceedings that are material to the business, meaning in this case that they involve greater than 10 percent of the company's assets. Environmental sanctions of $100,000 or more must also be disclosed.
- Item 303 requires discussion and analysis of environmental liabilities.

Under a 1990 agreement, the EPA provides the SEC with information about compliance with environmental laws, including names of parties receiving Superfund notice letters, cases filed under RCRA and Superfund, concluded federal civil environmental cases, criminal environmental cases, and names of RCRA facilities subject to corrective action. However, companies often report that environmental liabilities from these programs do not have a "material" effect on their business. They report that until the cleanup goal and time line for each of their contaminated sites are known, assessing the cost of cleanup is not possible.

---

vide large companies with a rational basis (reduced remediation liabilities on the corporate balance sheet) for measuring the value of remediation technologies. Recently, support has been growing for development of accounting practices that represent the full environmental costs of doing business, and consistent reporting of remediation liabilities would complement such full-cost accounting initiatives (see Box 2-8).

To provide for uniformity and credibility in corporate reporting of remediation liabilities, consistent standards for tabulating remediation liabilities would

---

BOX 2-8
Corporate Environmental Accounting

Support has been gathering over the past few years for the notion that full corporate reporting of environmental costs and liabilities is an essential element of business for economic sustainability and environmental stewardship (Ditz et al., 1995). Full-cost environmental accounting describes "how goods and services can be priced to reflect their true environmental costs, including production, use, recycling, and disposal" (Popoff and Buzelli, 1993). Thus, full-cost environmental accounting requires the inclusion of costs once considered to be external to corporate financial decisions. Ditz et al. (1995) refer to full-cost environmental accounting as the maintenance of "green ledgers."

Ditz et al. argue that "environmental costs are dispersed throughout most businesses and can appear long after decisions are made" primarily because "traditional accounting practices rarely illuminate environmental costs or stimulate better environmental performance." Environmental costs are often hidden and unrecognized in other categories. In case studies of several large corporations, Ditz et al. found that the true environmental costs of manufacturing operations were as high as 22 percent of operating costs. Full-cost environmental accounting and liability disclosure can help identify opportunities for cost savings that had not been previously recognized and exploited by managers.

---

need to be developed. The American Institute of Certified Public Accountants in October 1996 issued a position paper calling for more forthright reporting of environmental remediation liabilities, but detailed standards for such reporting do not yet exist (American Institute of Certified Public Accountants, 1996). The model used for federal taxation of corporations could be employed in developing remediation liability reporting standards. That is, a national body such as the Federal Accounting Standards Board could establish generally accepted accounting principles for such reporting. Third-party auditors (certified public accountants, ground water professionals, engineers, or all of these) could audit the records of the reporting company to ascertain whether the reports were accurate. Third-party auditing would greatly reduce the regulatory burden of monitoring compliance, in effect privatizing the audit function. The thoroughness and accuracy of the audits would, in turn, be established by creating third-party liability for auditors who fail to comply with accepted practices in preparing an audit. The potential for liability would lead to careful training and supervision of auditors. Public accountants are accustomed to the risk of being held liable for faulty audits, and they typically carry insurance against it.

The Geneva-based International Standards Organization (ISO) in late 1996 released a standard, known as ISO 14001, that prescribes how corporations can establish management systems, including accounting procedures, for keeping track of how all of a company's activities affect the environment (Begley, 1996). The ISO standard could serve as a model for developing a management process and an accounting system specific to keeping track of contaminated sites. The standard is expected to become widely used as a model for corporate accounting of environmental impacts (Begley, 1996). For example, the DOD and other U.S. agencies are conditionally requiring their vendors to become certified under the standard.

To eliminate financial disincentives for companies to comply with remediation liability reporting requirements, Congress should establish a remediation "mortgage" program. The program would allow a company to depreciate all of the remediation costs it declares at the outset of a project over a 20- to 50-year period, rather than having to subtract the full liability from its balance sheet all at once. The program would not be a true mortgage program because it would not involve financial lending. However, it would be similar to a mortgage program in that, rather than having to bear the full burden of remediation liability at once, companies could charge part of the cost against earnings each year, much as a homeowner generally has 15 to 30 years to pay off debt. Such a program would ensure that companies would not risk losing a major portion of their value by accurately and completely reporting all remediation costs they are likely to face in the coming decades. That is, companies would not have to bear the full impact of remediation liability at once. A remediation "mortgage" would have the added advantage of providing companies with a cost target (the cost of the "mortgage") to beat.

Although federal agencies must be accountable to the public and have a responsibility to spend tax dollars wisely, remediation at government sites is less driven by financial concerns than remediation at privately owned sites. At federal sites, financial resource allocations are driven more by goals and negotiated milestones than by costs (GAO, 1995b). Thus, the public-sector remediation market is less subject to influence by financial stimuli than the private-sector remediation market. Nevertheless, financial incentives for considering innovative remediation technologies could be created by careful oversight of remediation contractors. Rather than hiring contractors on a cost-reimbursable basis, federal agency managers should hire remediation contractors on a fixed-price basis, in which the cost of achieving a specified goal is agreed upon in advance and clear milestones are established. To provide assurance that remediation is proceeding toward those milestones in an efficient manner and protect against waste and abuse of government resources, federal site managers should establish independent peer review panels to check progress at specified milestones. Limiting the amount that can be charged toward remediation of federal sites and providing for independent review of progress at those sites would provide incentives for remediation contractors to implement efficient, innovative solutions. In some cases, site complexities will result in remediation costs much higher than those originally projected by the contractor. In such cases, the peer review panel could examine the request for a cost increase and determine whether it is technically justified.

Consistent Enforcement

A market that is a function of regulatory requirements as its core basis must at a minimum be consistent and predictable. For example, the U.S. tax system is based on self reporting, but it works because there are known and credible consequences for those who do not comply. Enforcement of waste site remediation requirements should be similarly consistent. Organizations will engage in remediation for two reasons: (1) because there is value in solving the problem or (2) because there are known negative consequences for noncompliance. It is imperative to have a predictable, known, and consistent enforcement mechanisms accompanied by high penalties. Without sufficient enforcement and penalties for noncompliance, the system rewards those who delay.

The financial resources and number of personnel dedicated to enforcement of waste site remediation regulations need to be increased so that those who do not comply are consistently penalized. In addition, enforcement penalties need to be higher than the costs of remediation to make noncompliance more costly than remediation. Third-party auditing of environmental liability, as described above, could be added to the existing set of regulatory enforcement tools. The EPA and the Department of Justice could pursue enforcement actions against companies whose audits reveal failure to comply with hazardous waste regulations.

Predictable Regulatory Requirements and Time Lines

The regulatory process for deciding on cleanup goals and selecting remediation technologies must be sufficiently uniform to justify the development cost of a new technology and thus leverage the cost over a wide group of customers. Consistency in the remedy selection process is not equivalent to establishing presumptive remedies that will be the preferred choice for cleaning up different types of contaminated sites. In fact, establishment of presumptive remedies runs counter to innovation by, in essence, freezing the menu of technologies at the point in time at which the presumptive remedies were developed. Rather, consistency in remedy selection processes means that the detailed steps in selecting remedies for two different sites having similar geophysical and contaminant characteristics should be similar, regardless of the regulatory program under which the sites are being cleaned up or the EPA office responsible for overseeing the sites. To increase the consistency of the remediation technology selection process, the EPA should conduct a detailed review of remedy selection procedures at Superfund and RCRA sites in its 10 regions. Based on this review, the EPA should determine the degree to which these procedures vary and should recommend how to make the process more consistent.

The EPA should also consider whether establishing national cleanup standards for ground water and soil would enhance the cleanup process by providing greater consistency. Such standards would be based on cancer and noncancer (such as neurological and reproductive) effects of contaminants, as well as ecosystem effects. They would need to include some mechanism to account for site-specific variations in the potential for human or ecosystem exposure to the contamination and in synergistic effects caused by the presence of multiple contaminants.

While national ground water and soil cleanup standards might benefit remediation technology developers by clarifying the level of performance that remediation technologies must achieve, the issue of whether such standards should be established is highly controversial and needs careful analysis. The Committee on Innovative Remediation Technologies could not reach consensus on whether such standards are advisable. Some members favored the establishment of standards because of the greater consistency they would provide and because such standards might create an incentive to achieve higher levels of cleanup, much as the establishment of standards for drinking water has spurred development of improved water treatment technologies. However, other members objected to recommending the establishment of national standards because they believe such standards might limit opportunities for site-specific judgment of appropriate cleanup levels by trained professionals. Nonetheless, the committee did agree that the issue of whether national ground water and soil cleanup standards should be established warrants careful consideration. Many states already have state-wide cleanup standards for soil, ground water, or both. If national standards were developed, site-specific assessment could always be an alternative and may be more appropriate for large, complex sites. The EPA and the Congress should review the effectiveness of state cleanup standards and the rationale for establishing them and determine whether national standards for soil and ground water cleanup would help advance the state of development of cost-effective subsurface remediation technologies.

As part of this effort, the EPA should also establish guidelines that would indicate tentative time lines for reaching the various regulatory milestones (site investigation, remedy selection, remedy construction) at sites with varying degrees of complexity, with more complex sites having longer remediation time lines. Site-specific flexibility is essential to allow for more detailed studies and longer time lines where initial investigations reveal site complexities. Nonetheless, general guidance on remediation time lines based on site complexity would help technology developers anticipate with greater certainty how long they might have to wait before receiving a job contract. Although the EPA prepares quarterly management reports that document the average duration of stages in the process of cleaning up Superfund sites, such averages are of limited use to remediation technology vendors because of the wide deviations from these averages and because the averages do not apply to cleanup efforts occurring outside of the Superfund program. To the extent possible, state-run remediation programs should follow the EPA's guidelines and general remedy selection processes.

To further increase the predictability of remediation time lines at Superfund sites, steps should be taken to reduce litigation associated with identifying potentially responsible parties. According to the GAO (1994b), factors that can help decrease the amount of litigation over who should pay for cleanup include careful work by the EPA to identify all potentially responsible parties up front, consistent enforcement against responsible parties who fail to meet regulatory requirements, and involvement of skilled mediators with the full group of responsible parties to negotiate their individual responsibilities.

Freedom to Choose

To provide incentives for innovation, customers must have the freedom to choose any remediation technology or group of technologies they desire in order to meet the required cleanup standards. Theoretically, regulators should be indifferent about how a company or federal agency cleans up a site, as long as the regulatory requirements for risk reduction are met. Current regulatory preapproval of remediation technologies should be curtailed. At the same time, the public will need assurance that in allowing this freedom, the public's goals for remediation are still achieved. Table 2-5 shows how companies could be allowed freedom to choose remediation technologies while still providing assurance to the public that cleanup standards will be achieved.

TABLE 2-5 Elements of a Regulatory System that Allows Freedom to Choose Remediation Technologies

Industry Goals	Public Safeguards
Performance-based regulation	No increase in risk (no backsliding on standards); increased penalties and liability for noncompliance
Flexibility in choosing treatment technology, including ability to change technology if a better alternative emerges	Prior public notice of remediation plan, with full disclosure of contamination conditions
Confidentiality of proprietary data	Auditing of data by third parties
Reduced cost	Continuing right to litigate if standards are not met

Massachusetts is pioneering a program for allowing customer freedom to choose remediation technologies that could serve as a useful model at the federal level (Huang, 1995). In Massachusetts, consultants can apply to become "licensed site professionals" who can select remediation technologies, without regulatory approval, for sites where cleanup is required under the Massachusetts Contingency Plan. Licensed site professionals must have 8 years of experience in hazardous waste consulting, with 5 of those years consisting of experience as a principal decisionmaker, and their qualifications must be validated by a state review board. They can approve remedies for all but a few sites that the state has determined are highest priority. State regulators audit about 20 percent of the sites each year to ensure that licensed site professionals are complying with regulatory cleanup requirements. One problem with this system has been that licensed site professionals can hesitate to choose innovative technologies because of the fear that they will not work. To help overcome this fear of risk, Massachusetts has developed guidance documents on innovative technologies for licensed site professionals, an on-line data base with innovative technology performance information, and educational sessions focusing on innovative technologies. In addition, the state provides regulatory incentives such as reduced fees and extended deadlines when innovative technologies are used. Although licensed site professionals have been somewhat conservative in selecting technologies, the system nonetheless has added certainty to remediation of hazardous waste sites in Massachusetts. Companies know that once a licensed site professional selects a remedy for their site, they can implement cleanup without fear of regulatory delays.

Full Disclosure of Contaminated Sites

Companies, as well as government agencies, should be required to fully disclose information about all contaminated sites above a given size or risk level. Included in this disclosure should be descriptions of contaminants present at the site, geologic conditions, and releases into the soil, ground water, and air. Environmental impacts and risks to public health, wildlife, and ecosystems should also be evaluated and publicly disclosed. Such a public disclosure requirement would provide technology developers with information about the size and nature of the remediation market. It would also increase awareness of contamination problems among stakeholders and thus provide incentives and support for prompt site remediation. Already, some state laws, such as New Jersey's Environmental Cleanup Responsibility Act, require disclosing the environmental condition of a site when it is being transferred to another party. A national disclosure program would go beyond such state programs in that it would not require property transfer as a trigger. While there is political pressure to avoid including sites on registries such as the Superfund National Priorities List because of the perceived stigma associated with having a site on such a list, public disclosure of contaminated site information is essential not only for the benefit of communities potentially affected by these sites but also to clarify market opportunities for remediation technology developers.

The EPA could be responsible for compiling all of the information from the site disclosure program, as well as information from abandoned sites requiring cleanup, into a national registry that could be included in a home page on the Internet. The registry should include all types of contaminated sites, including those governed by the Superfund program, the RCRA program, and state programs. New York has a hazardous waste site registry consisting of concise site reports that might serve as a model for a national registry. The information in the registry should be indexed according to location, owner/operator, site characteristics, contaminant types, off-site impacts, and regulatory status. The new site inventory would be analogous to the existing Toxics Release Inventory (TRI) created under the Emergency Planning and Community Right-to-Know Act of 1986, which requires industries to report certain releases to air, water, and land. The TRI reporting requirement has had the unintended but beneficial effect that companies have learned of emissions about which they were previously unaware and have found ways to reduce or eliminate them altogether, in part in order to avoid standing out as the worst polluter. In addition to creating an essential source of market data for technology developers, a requirement for full disclosure of remediation-related liabilities would provide companies and agencies with more data about the impacts of the contaminated sites, regulators with better information for analyzing associated risks and alternatives, and the public with a better basis for involvement in the selection of remedies. Disclosure would help ensure corporate and regulatory accountability for remediation decisions.

Technology Demonstration and Verification

Given the hesitancy of corporations to serve as the first client for an innovative remediation technology, more opportunities need to be created to test innovative technologies and verify their performance prior to marketing. The EPA has recognized this problem and has in place initiatives to encourage the testing of innovative remediation technologies at federal facilities, including DOD and DOE sites, as described in detail in Chapter 5. Such programs should be given a high priority. If testing on a federal facility proves the technology is effective and cost competitive, the government should guarantee that it will use the technology at least once at a federal facility.

Also needed is a coordinated program for formally verifying remediation technology performance. Official, federally sanctioned verification of technology performance provides customers with assurance that performance data on new technologies are valid and representative of the future expected performance of the technology. Performance verification could also reduce regulatory barriers and hence time to market entry and facilitate the raising of capital needed to commercialize new technologies. As explained in Chapter 5, existing technology demonstration and verification programs are uncoordinated and lack credibility across market sectors. Chapter 5 explains the details of how technology performance can be verified.

CONCLUSIONS

Increasing the use of innovative remediation technologies at public- and private-sector sites will require a shift in the paradigm that currently governs the remediation market. Rather than being driven by environmental regulations alone, organizations responsible for contaminated sites need to be motivated to pursue remediation for financial reasons. Making this transition to a market-oriented system for remediation will require that environmental regulators allow organizations with contaminated sites the freedom to choose how they will accomplish the required remediation end points; it will require organizations responsible for contamination to honestly evaluate and disclose the full costs of site remediation.

Shifting to a market-oriented approach to contaminated site remediation would create incentives for faster and more effective subsurface cleanups and would thus revitalize the market for remediation technologies. Instead of searching for ways to delay cleanup, organizations would be prompted by financial self interest to expedite remediation. A market-based approach would also allow for more efficient allocation of corporate and regulatory resources. Regulators could shift their attention from organizations that are actively cleaning up their sites to those that are lagging. Companies working to meet the regulatory requirements would benefit from reduced bureaucratic transaction costs. To date, the market for remediation technologies has been constrained by lack of customer demand, not because the number of contaminated sites is small but because customers have failed to perceive remediation as an activity undertaken in their economic self interest, rather than as a pure expense.

The rest of this report focuses on technological solutions to the problems of commercializing innovative remediation technologies. However, these technological initiatives will be ineffective without concurrently stimulating the market forces needed to create demand for better, less costly remediation technologies.

RECOMMENDATIONS

To amplify the market forces for remediation technology commercialization, the following steps should be taken:

• **The SEC should clarify and strictly enforce requirements for disclosure of environmental remediation liabilities by all publicly traded U.S. corporations.** Clarifying the existing requirements for reporting of environmental liabilities and strictly enforcing these requirements would provide an incentive for companies to initiate remediation, rather than delaying it, in order to clear their balance sheets of this liability. Detailed accounting procedures for complying with this requirement, along with a mechanism for certifying environmental accountants, need to be established by the U.S. accounting profession, possibly using the model of the International Standards Organization's series of standards for environmental management systems. Although technical uncertainties will preclude exact computations of remediation liabilities, companies should nonetheless be required to report their best estimates of these liabilities using reasonable estimates of probable remediation scenarios.

• **The SEC should enforce environmental liability reporting requirements through a program of third-party environmental auditing.** The possibility of an environmental audit, along with strong penalties for failing the audit, would help ensure that companies would comply with SEC requirements to report environmental liabilities. Certified public accountants, ground water professionals, or all of these could conduct the audits after receiving appropriate training.

• **Congress should establish a program that would allow companies to amortize the remediation liabilities they report over a 20- to 50-year period.** Such a program would ensure that by fully evaluating and disclosing their remediation liabilities with the best available current information, companies would not risk losing a major portion of their value. It would also provide a measurable cost target for remediation technologies to beat (the total cost of the declared liabilities).

• **The EPA should work to improve enforcement of Superfund and RCRA requirements.** Consistent, even-handed enforcement is essential for ensuring that U.S. companies are not placed at a competitive disadvantage compared to their domestic competitors by spending money on remediation.

• **Managers of federal hazardous waste sites should hire remediation contractors on a fixed-price basis and should establish independent peer review panels to check progress toward specified remediation milestones.** Such steps are necessary to provide stronger incentives for federal remediation contractors to implement efficient, innovative solutions to contamination problems. When site complexities result in remediation costs that exceed the initial estimates, the peer review panel could verify that the cost increase is technically justified.

• **The EPA should review procedures for approving remediation technologies in its 10 regions and should develop guidelines for increasing the consistency and predictability of these procedures among regions and across programs; to the extent possible, state hazardous waste remediation programs should follow these guidelines.** A consistent regulatory process that responds rapidly to approval requests is essential so that remediation technology developers can predict with reasonable certainty the steps that will be required for regulatory approval of their technology and how long they may have to wait before receiving their first job contract. While the process for remedy selection should be the same at each site, site managers must have the flexibility to consider any remediation technology that they believe will meet regulations at the lowest possible cost, provided the public has sufficient opportunity to voice concerns during the remedy selection process and to challenge the selected remedy.

• **Congress and the EPA should assess the arguments for and against establishing national standards for ground water and soil cleanup.** While some states are adopting state-wide cleanup standards, no national standards exist. Such standards might increase the predictability of the remediation process and consistency in the approaches used in the many remediation programs; predictability and consistency would benefit technology developers by providing them with a more certain end point for remediation. On the other hand, such standards might have the detrimental effect of decreasing flexibility in site remediation. The issue of whether national cleanup standards are advisable should be carefully considered.

• **The GAO should investigate the Massachusetts program for licensing site professionals to select remediation technologies on behalf of environmental regulators and should recommend whether such a program should be implemented nationally.** Such a program might help to eliminate delays associated with regulatory approval steps.

• **The EPA should establish a national registry of contaminated sites similar to the Toxics Release Inventory and should make it publicly available on the Internet.** Such a registry would allow technology developers to assess the size and characteristics of different segments of the remediation market. It would also provide an incentive for companies to clean up sites quickly in order to remove them from the registry. Although there is political pressure to avoid including contaminated sites on registries because of the perceived stigma associated with owning a site on such a list, public disclosure of contaminated site information is essential for ensuring that accurate and complete information about the remediation market is widely available.

• **Federal agencies should continue to support and expand programs for testing innovative remediation technologies at federal facilities.** Providing opportunities for testing full-scale technology applications is essential for new technology ventures that need cost and performance data to provide to potential clients.

REFERENCES

American Institute of Certified Public Accountants. 1996. Statement of Position 96-1: Environmental Liability. Jersey City, N.J.: American Institute of Certified Public Accountants.

Begley, R. 1996. ISO 14000: A step toward industry self-regulation. Environmental Science & Technology 30(7):298A-302A.

Bianchi-Mosquera, G. C., R. M. Allen-King, and D. M. Mackay. 1994. Enhanced degradation of dissolved benzene and toluene using a solid oxygen-releasing compound. Ground Water Monitoring and Remediation 9(1):120-128.

Blackwelder, B. 1996. Disclosing environmental liability to shareholders. Shareholder proposal presented at the Eastman Kodak shareholder meeting, May 8, Rochester, N.Y.

Conestoga-Rovers & Associates. 1986. Surface Soil Sampling and Analysis Program/Royalton-Hartland and Gasport School Properties/Middleport, N.Y. Waterloo, Ontario: Conestoga-Rovers & Associates.

Congressional Budget Office. 1994a. Analyzing the Duration of Cleanup at Sites on Superfund's National Priorities List. Washington, D.C.: Congressional Budget Office.

Congressional Budget Office. 1994b. The Total Costs of Cleaning Up Nonfederal Superfund Sites. Washington, D.C.: U.S. Government Printing Office.

Ditz, D., J. Ranganathan, and R. D. Banks. 1995. Green Ledgers: Case Studies in Corporate Environmental Accounting. Washington, D.C.: World Resources Institute.

Dixon, L. S., D. S. Drezner, and J. K. Hemmitt. 1993. Private-Sector Cleanup Expenditures and Transaction Costs at 18 Superfund Sites. Santa Monica, Calif.: RAND.

ENR (Engineering News Record). 1996a. Wichita works it out. ENR (July 15):36-37.

ENR. 1996b. Cleanup firms getting realistic. ENR (October 21):16.

EPA. 1995a. Geosafe Corporation: In Situ Vitrification—Innovative Technology Evaluation Report. EPA/540/R-94/520. Washington, D.C.: EPA, Office of Research and Development.

EPA. 1995b. Market Opportunities for Innovative Site Cleanup Technologies: Middle-Atlantic States. EPA-542-R-95-010. Washington, D.C.: EPA, Office of Solid Waste and Emergency Response.

EPA. 1996a. Innovative Treatment Technologies: Annual Status Report, Eighth Edition. EPA-542-R-96-010. Washington, D.C.: EPA, Office of Solid Waste and Emergency Response.

EPA. 1996b. Market Opportunities for Innovative Site Cleanup Technologies: Southeastern States. EPA-542-R-96-007. Washington, D.C.: EPA, Office of Solid Waste and Emergency Response.

FMC. 1984. NYS Hazardous Waste Disposal Questionnaire to Langdon Marsh, NYS DEC Deputy Commissioner, filed August 24, 1984. Albany, N.Y.: Department of Environmental Conservation.

FMC. 1995. Chronology of Pertinent Environmental Issues. Middleport, N.Y.: FMC.

GAO (General Accounting Office). 1994a. Nuclear Cleanups: Difficulties in Coordinating Activities Under Two Environmental Laws. GAO/RCED-95-66. Washington, D.C.: GAO.

GAO. 1994b. Superfund: Legal Expenses for Cleanup-Related Activities of Major U.S. Corporations. GAO/RCED-95-46. Washington, D.C.: GAO.

GAO. 1995a. Community Development: Reuse of Urban Industrial Sites. GAO/RCED-95-172. Washington, D.C.: GAO.

GAO. 1995b. Federal Hazardous Waste Sites: Opportunities for More Cost-Effective Cleanups. GAO/T-RCED-95-188. Washington, D.C.: GAO.

GAO. 1996a. Superfund: Barriers to Brownfield Redevelopment. GAO/RCED-96-125. Washington, D.C.: GAO.

GAO. 1996b. Superfund: How States Establish and Apply Environmental Standards When Cleaning Up Sites. GAO/RCED-96-70. Washington, D.C.: GAO.

Guerrero, P. F. 1997. Superfund: Times to assess and clean up hazardous waste sites exceed program goals. Testimony before the Subcommittee on National Economic Growth, Natural Resources, and Regulatory Affairs, Committee on Government Reform and Oversight, U.S. House of Representatives, February 13, 1997. GAO/T-RCED-97-69. Washington, D.C.: GAO.

Hembra, R. L. 1992. Superfund: Problems with the completeness and consistency of site cleanup plans. Testimony before the Subcommittee on Investigations and Oversight, Committee on Public Works and Transportation, U.S. House of Representatives, June 30, 1992. GAO/T-RCED-92-70. Washington, D.C.: GAO.

Houlihan, J. 1995. Presentation to the National Research Council's Committee on Innovative Remediation Technologies, National Academy of Sciences, Arnold and Mabel Beckman Center, Irvine, California, March 9.

Huang, R. 1995. Presentation to the National Research Council's Committee on Innovative Remediation Technologies, National Academy of Sciences, J. Erik Jonsson Woods Hole Center, Woods Hole, Massachusetts, May 25.

Jarosch, T. R., R. D. Raymond, and S. A. Burdick. 1995. Sampling and Analysis Report: Thermatrix Flameless Thermal Oxidation Field Demonstration at the Savannah River Site. Prepared for Lockheed Martin Energy Systems, Inc., Hazardous Waste Remedial Actions Program. Aiken, S.C.: Westinghouse Savannah River Company, Savannah River Technology Center.

National Research Council. 1994. Alternatives for Ground Water Cleanup. Washington, D.C.: National Academy Press.

NYS DEC (New York State Department of Environmental Conservation). 1987. Surface and Subsurface Soil/Sediment Investigations at Royalton-Hartland Schoolyard, Jeddo Creek, Culvert 105 Extension. Albany, N.Y.: Department of Environmental Conservation.

NYS DEC. 1996. FMC—Middleport Facility Groundwater Monitoring and Off-Site Soil Sampling Issues Correspondence File 4/12/95 to 5/9/96. Albany, N.Y.: Department of Environmental Conservation, RCRA Division.

NYS DOH (New York State Department of Health). 1987. Results of Biological Monitoring Program for Arsenic and Lead: Middleport Elementary and Roy-Hart Jr./Sr. High Schools. Albany, N.Y.: Department of Health, Bureau of Environmental Epidemiology and Occupational Health.

Olsen, R. L. 1996. Alternative cleanup criteria and innovative approaches for groundwater remediation. Pp. 305-314 in Proceedings, Hazwaste World Superfund XVII. Bethesda, Md.: E. J. Krause and Associates.

Popoff, F., and D. Buzelli. 1993. Full-cost accounting. Chemical and Engineering News 77(2):8-10.

Renner, R. 1996. DOE Environmental Management Program awards first basic research grants. Environmental Science & Technology 30(10):431A.

Robb, K. E. B. 1993. SEC trains its eye on the environment. National Law Journal (March 8):23.

Roberts, R. 1994. Environmental reporting and public accountability. Paper presented at the Environmental Law Institute Corporate Management Workshop, Washington, D.C., June 6, 1994.

Roberts, R. Y., and K. R. Hohl. 1994. Environmental liability disclosure and Staff Accounting Bulletin No. 92. The Business Lawyer 50(November):1.

Russell, M., E. W. Colglazier, and M. R. English. 1991. Hazardous Waste Remediation: The Task Ahead. Knoxville: University of Tennessee, Waste Management Research and Education Institute.

Schofield, J. 1995. Presentation to the National Research Council's Committee on Innovative Remediation Technologies, National Academy of Sciences, Arnold and Mabel Beckman Center, Irvine, California, March 9.

SBA (Small Business Administration). 1994. Bridging the Valley of Death: Financing Technology for a Sustainable Future. Washington, D.C.: SBA.

SEC (Securities and Exchange Commission). 1993. Staff Accounting Bulletin No. 92. Washington, D.C.: SEC.

Stone, W. 1975. Memorandum from Ward Stone, wildlife pathologist, NYS DEC, to Ken Cohen, U.S. Attorney, July 14, 1995.

Value Line, Inc. 1996. The Value Line Investment Survey, Volume LI, No. 41. New York: Value Line Publishing Co., Inc.

VentureOne Corp. 1996. 1995 Annual Report. San Francisco: VentureOne Corporation.

3

State of the Practice of
Ground Water and Soil Remediation

Innovation in the environmental industry is driven by the need to solve difficult problems and the desire to improve upon existing solutions. When a ground water or soil cleanup technology is developed and applied, frequently in response to an unsolved problem, its acceptance and application are often limited initially to specific contaminants and specific hydrogeologic conditions. As the technology matures, it typically addresses the same range of contaminant types, but its range of application in subsurface environments becomes better defined. This evolutionary process is similar for most remediation technologies, but the rate at which new technologies are adopted varies considerably. For example, soil vapor extraction (SVE) technologies, used for removing volatile contaminants from soil, were virtually unused at Superfund sites in 1985 but by 1995 had been selected for source control at 20 percent of Superfund sites (EPA, 1996a). However, for other technologies, especially those for cleaning up contaminants in situ, this evolution is occurring much more slowly than one would predict based on the large number of contaminated sites and the hundreds of billions of dollars in projected cleanup costs for these sites. There is no shortage of new ideas for improving the ability to restore contaminated ground water and soil. However, for reasons explained in Chapter 2, successful commercialization of all but a few new ideas has been limited.

This chapter reviews the state of development of technologies for cleaning up ground water and soil, highlighting knowledge and information gaps, and describes challenges and strategies for cleaning up different types of contaminants. The chapter defines all technologies for cleaning up contaminants below the water table as "ground water cleanup technologies" and all technologies for cleaning up contaminants above the water table as "soil cleanup technologies." This distinction is somewhat artificial, because many technologies for restoring areas below the water table address contaminated geologic materials rather than the water itself. Nevertheless, although these technologies do not specifically treat the water, but rather contaminants in the geologic materials, users of the technologies generally refer to them as ground water cleanup technologies because their primary intent is to prevent the contaminants from dissolving in and contaminating the ground water.

Included in this chapter are technologies that treat ground water contaminants in place in the subsurface and soil technologies that treat the soil either in place or on site in a treatment unit. The chapter does not cover technologies for removing contaminants from ground water once it has been pumped to the surface. The challenge of removing contaminants from water at the surface has already been largely addressed through the development of systems for treating municipal and industrial wastewater. In comparison, relatively few technologies are available for removing contaminants from soil or geologic materials to which the contaminants have tightly bound. Even fewer technologies exist for treating contaminated ground water in place in the subsurface. Furthermore, the processes that can be exploited in these technologies are still not fully understood.

WHAT IS INNOVATIVE REMEDIATION TECHNOLOGY?

"Innovative technology" as applied to the cleanup of ground water and soil is an elusive term, for two primary reasons. First, government agency representatives and others involved in waste-site cleanup may have different perspectives on which technologies are innovative. For example, the Environmental Protection Agency's (EPA's) 1996 *Innovative Treatment Technologies: Annual Status Report* classifies in situ bioremediation of contaminated soils as an innovative technology, while the Air Force specifies bioventing (a type of in situ bioremediation) as the standard remedy for soils contaminated with petroleum hydrocarbons and other volatile organic compounds (DOD Environmental Technology Transfer Committee, 1994). The Department of Energy (DOE) considers any technology innovative if it has not been used at DOE sites (J. Walker, DOE, personal communication, 1995). Thus, the definition of innovative varies depending on the perspective of the user.

A second reason why "innovative" is hard to define is that technologies are continually evolving. In the ground water and soil remediation business, only a few technologies represent true breakthroughs, in the sense that they apply concepts never before used in the field. More commonly, innovation occurs incrementally, evolving from existing technologies. This evolution is a product of several factors. The first factor is increased experience. As a technology is implemented at various sites, the experience gained provides a basis for defining and overcoming limitations, establishing best practices, and expanding the technology's application to new contamination problems. The second factor is competition, both among the technologies and among practitioners who design the technologies. The third factor is the technology's performance limitations. All technologies have practical limitations that affect their market viability; a given technology may not perform well initially for certain contaminant types and hydrogeologic settings, but as these limitations are addressed, the applicability and marketability of the technology may increase. The fourth factor is cross-fertilization with other technologies. A technique or even a whole new technology may be incorporated into an existing technology to improve its performance or overcome a limitation.

The net result of this evolution is that remediation technologies go through a cyclical or generational life cycle. The first stage is the initial, and often crude, application of the technology. During this period, acceptance of the technology increases as it is successfully applied and its success is communicated. During the second stage of evolution, design practices for the technology become established. Acceptance and application grow rapidly as the benefit of the technology becomes known. The third stage is the mature, common practice of the technology. In this stage, the focus shifts from the benefits of the technology to its limitations, and acceptance and application may decline. During this mature application stage, the technology is vulnerable to replacement. However, use of the technology may increase again, either as its limitations are overcome or as the technology becomes applicable to new contaminant types and/or hydrogeologic settings. When a significant limitation is overcome, the technology enjoys a rebirth. The two best examples of remediation technologies that have developed through this evolutionary process are in situ bioremediation and SVE (see Boxes 3-1 and 3-2).

This chapter reviews a broad range of remediation technologies other than those based on conventional pumping and treating of ground water or digging and either hauling or burning of soil. Many of the technologies discussed in the chapter, including SVE and in situ bioremediation, have a significant experience base and are not new. In addition, many are enhancements to conventional approaches rather than new developments. Nevertheless, all of these technologies have in common the ability (whether potential or proven) to increase the effectiveness and/or decrease the costs of subsurface cleanup when compared to the historical approaches of pumping and treating ground water and hauling or burning soil.

AVAILABILITY OF INFORMATION ON INNOVATIVE REMEDIATION TECHNOLOGIES

Application of innovative remediation technologies has been slowed by lack of uniform, synthesized information about remediation technology performance. While considerable effort is being invested in researching and developing remediation technologies, these efforts are often isolated and do not benefit the general industry because circulation of information is limited. Broad acceptance of a tech-

BOX 3-1
Innovations in Engineered In Situ Bioremediation

Engineered in situ bioremediation, the purposeful stimulation of microorganisms in ground water and soil to degrade contaminants, was first applied in 1972 to clean up a spill of gasoline from a pipeline in Ambler, Pennsylvania (Raymond et al., 1977). At that time, in situ bioremediation addressed the unsolved problem of residual petroleum contamination in soil and ground water. The related generations of technology that followed this first in situ bioremediation system focused primarily on petroleum hydrocarbons and were essentially water-based systems: both the oxygen and nutrients necessary to stimulate growth of contaminant-consuming organisms were supplied by circulating ground water. From 1972 to 1983, only about a dozen in situ bioremediation projects were conducted nationwide (Brown et al., 1993). Performance was limited by the low solubility of oxygen in water.

The second generation of in situ bioremediation systems used hydrogen peroxide (H_2O_2) to provide a more efficient method of supplying oxygen in soluble form. H_2O_2 supplied 10 to 50 times more oxygen equivalents than did the existing aeration systems. The third-generation in situ bioremediation system employed SVE (see Box 3-2) to supply oxygen above the water table and used H_2O_2 to provide oxygen below the water table. Both the second- and third-generation technologies were limited by the expense and difficulty of using H_2O_2, and there was a significant effort to find an alternative. Several alternatives to H_2O_2 were explored, but the most successful was air sparging, which involves injecting air directly into the subsurface. Air sparging was initially developed as a separate technology to remove volatile contaminants by evaporative processes, but it was soon incorporated into bioremediation technology, sometimes called biosparging. Due in part to the success in improving oxygen delivery systems, in situ bioremediation systems are now in use at thousands of underground storage tank sites and dozens of Superfund sites (see Chapter 1).

Parallel to the development of the different generations of in situ bioremediation systems for treatment of petroleum hydrocarbons has been the development of a number of spin-off technologies. The first of these was improved ex situ soil bioremediation technology. Ex situ bioremediation employs the principles and techniques of in situ bioremediation to treat excavated soils. A second improvement has been the increased understanding of bioremediation of chlorinated hydrocarbons, which was largely unknown until the mid-1980s (McCarty and Semprini, 1994). A third spin-off has been the development of intrinsic bioremedi-ation (bioremediation without using engineered systems to stimulate native soil microbes) to control and mitigate contaminant plumes.

Petroleum-contaminated soils being treated in above-ground bioremediation cells engineered with vapor extraction systems. Courtesy of Fluor Daniel GTI.

nology requires documentation of the technology's performance and accessibility of performance information. Often in the development of remediation technology, data collection is minimal. As a result, much of the available information is anecdotal and empirical. This relative lack of documented performance data makes it difficult to judge the benefits and limitations of a technology without trying it. As a result, remediation technology development is somewhat repetitive, as individual practitioners tend to repeat the same work until the experience base is sufficiently distributed that knowledge of the technology is also well distributed.

The lack of consistent information is pervasive in the remediation market and encompasses all of the following problems:

- technology reports are often incomplete;
- critical scientific evaluation of technology application most often is not conducted or is not conducted with the goal of collecting comparable data sets;
- reliable cost data are lacking and inconsistent;
- methods for determining costs and evaluating successes can vary enormously; and
- much information is proprietary.

These factors combine to make it very difficult to conduct rigorous comparisons between technologies and across problem contexts with existing data. In addition, there are no complete, centralized data bases that cross markets and government programs. The 1995 publication *Accessing Federal Data Bases for Contaminated Site Clean-Up Technologies* lists 25 different data bases that could be potentially useful in evaluating remediation technologies (Federal Remediation Technologies Roundtable, 1995a), but these data bases are not coordinated, and many of them are difficult to access (see Appendix A for a listing of data bases). In fact, the existence of such a large number of data bases in itself creates confusion, because the data bases contain information in different formats that may not be comparable, and the quality of the data from different data bases is variable and difficult to assess.

A few programs exist or are developing to facilitate evaluation of technologies. The EPA's Technology Innovation Office is collecting information on Superfund technology selections. The Federal Remediation Technologies Roundtable (a consortium of federal agency personnel involved in remediation) has developed protocols for standardized cost evaluations (Federal Remediation Technologies Roundtable, 1995b). Joint programs between states allowing sharing of information collected for technology evaluation are being created, and the results from these programs should help alleviate some of the information deficit. The EPA has established the Ground Water Remediation Technologies Analysis Center to help disseminate information on new remediation technologies (GWRTAC, 1995). These efforts are useful for developing a global view of what categories of remediation technologies are being tested and implemented. However, the fact remains that there are few reports that contain well considered evaluations of technologies, rather than mere compilations of information lacking careful analysis. A few examples of thoughtful technology evaluation may serve as models for future work.

The American Academy of Environmental Engineers' WASTECH® project has produced eight monographs of innovative waste-site remediation technologies. Generally, the scope of the analysis for this project was limited to technologies that are not commonly applied; have been sufficiently developed so that they can be used in full-scale applications; have sufficient data available to describe and explain the technology; and have sufficient data to assess effectiveness, limitations, and potential applications. An important contribution of this effort was applying such criteria to a wealth of information and synthesizing the findings by a task group of experts, whose work was peer reviewed, to produce a discussion of potential applications, process evaluations, limitations, and technology prognoses. Although restricted in scope and not uniform in coverage, the WASTECH® monographs provide a measure of consensus on performance of the technologies reviewed in the series; the philosophy and approach are laudable. A follow-up WASTECH® series of seven monographs emphasizing remediation technology design and implementation is in preparation.

An example of a focused report on a particular problem context is Dávila et

---

BOX 3-2
Innovations in Soil Vapor Extraction (SVE)

SVE has been used to treat volatile hydrocarbons since the mid-1970s. Originally, the technology was used to remove vapors from soils to prevent the vapors from entering buildings. This first-generation technology was derived from methane collection systems employed at landfills and typically consisted of lateral collection pipes placed along building foundations. A vacuum was applied to these lateral pipes to collect the organic vapors. Engineers soon observed, however, that the removal of vapors led to significant contaminant mass removal and a reduction of the level of contamination in the soil.

The second-generation (circa 1983) SVE technology focused specifically on the removal of volatile organic compounds (VOCs) from soil instead of the simpler collection of vapors. Two theories developed concerning SVE. The first postulated that the function of the vacuum was to "vacuum distill" the VOCs from the soil matrix and was based on the principle that the boiling point of most VOCs decreases with decreasing pressure. With this approach, typically a high vacuum (greater than 500 mm, or 20 in., Hg) was applied to decrease the boiling point of the VOCs and allow them to volatilize. The second theory of SVE postulated that the process was an evaporative one. The purpose of the vacuum was to induce air flow through the subsurface to evaporate the VOCs. This theory has become the dominant SVE theory and is the basis for most SVE designs. The evaporative model uses low to moderate vacuums of less than 380 mm (15 in.) Hg.

The main evolution of SVE has been in the design tools. The second-generation systems were typically designed on the basis of vacuum radius. Designers assumed that as long as there was a detectable vacuum,

---

al. (1993), which discusses technology applications for cleanup of soil and sediment contaminated with polychlorinated biphenyls (PCBs). This report discusses succinctly a number of technologies from an engineering perspective, pointing to specific examples of successes and problems from field evaluations. A report by Grubb and Sitar (1994) that discusses in situ remediation of dense nonaqueous-phase liquid (DNAPL) contaminants provides another example of a report that critically evaluates technologies for solving a particular type of contamination problem. A report by Troxler et al. (1992) on thermal desorption for petroleum-contaminated soils provides a useful perspective on a particular technology application and its status of development. Reports by Vidic and Pohland (1996) on in situ treatment walls and by Jafvert (1996) on cosolvent and surfactant flushing systems provide peer-reviewed evaluations of these technologies.

there was sufficient air flow. With this simple design basis, however, many SVE systems proved less effective than planned. Designs based on simple vacuum readings do not reflect true air flow unless they are adjusted for air permeability using Darcy's law. Once models were used to adjust vacuum readings to actual air flow rates, the performance of SVE systems improved. In retrospect, use of air flow-based designs seems like an obvious improvement. However, implementation has not been easy because of the need to use flow models.

While SVE has not evolved as extensively as bioremediation, it has engendered a much more varied set of spin-off technologies. This may be in part due to the clarity of the limitations of SVE, which removes volatile compounds from unsaturated soils by induced air flow. By this definition, there are three basic limitations: (1) the volatility of the contaminant, (2) the lack of air in saturated environments, and (3) the permeabilty of the soil matrix to air flow.

The limitation of volatility has fostered two other innovations. The first is the use of thermal energy to increase the contaminant volatility. Thermal systems under development use hot air, steam, radio waves, microwaves, or electrical resistance to heat the soil. The second is the use of biodegradation to enhance contaminant removal. While SVE was recognized as an efficient source of oxygen for bioremediation as early as 1984, the full development of bioventing did not occur until about 1990.

The lack of air in the saturated zone has fostered the development of dual-phase technology. This is the direct use of a high vacuum to dewater and vent saturated soils. Dual-phase technology has allowed SVE to treat contamination below the water table. Finally, the lack of air permeability has led to the development of fracturing systems. Fracturing systems inject pressurized air or water to open channels in the soil, which then allow air to circulate more freely.

STATE OF INNOVATIVE REMEDIATION TECHNOLOGY DEVELOPMENT

The current state of remediation technology development is relatively rudimentary. That is, technologies are available for treating easily solved contamination problems—mobile and reactive contaminants in permeable and relatively homogeneous geologic settings—but few technologies are available for treating recalcitrant contaminants in complex geologic settings. Figure 3-1 shows a conceptual diagram of where innovation is most needed to improve the performance and reduce the costs of ground water and soil remediation projects.

The greatest successes in remediation to date have been in the treatment of petroleum hydrocarbon fuels—gasoline, diesel, and jet fuel—which are generally mobile and biologically reactive and to a lesser extent in the treatment of chlori-

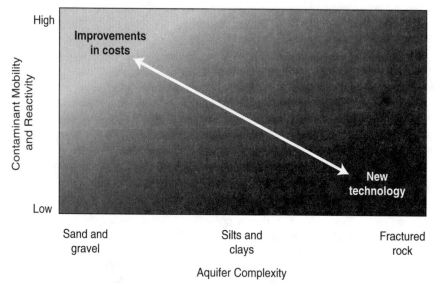

FIGURE 3-1 Technology needs for remediation of contaminated ground water and soil. At the left side of the figure, improvements are needed primarily to reduce costs. At the right side, new technologies are needed to solve contamination problems that are currently intractable.

nated solvents, which are generally mobile but are less readily biodegraded than petroleum hydrocarbons. The greatest challenge in remediation is in the location and cleanup of contaminant source material. This source material may comprise organic solids, liquids, or vapors; inorganic sludges and other solid-matrix wastes; compounds adsorbed on mineral surfaces; and compounds adsorbed in natural organic matter such as humus. Often, contaminant sources are difficult to locate and delineate because of lack of information about the contaminant spill or disposal history at the site and because contaminant source material may migrate away from where it was originally lost to the environment. Once found, source material may be inaccessible, lying under structures, or at great depth, or in fractured rock. Because of the possibility of continual contaminant release, partial source removal may not result in a proportional increase in ground water quality. The time for source diminution may be excessively long. Further, directing pumped fluids to the region where such fluids are most beneficial may be very difficult because of the problem of preferential flow, which causes fluids to bypass altogether the less permeable regions containing contaminants. A special challenge in the cleanup of source material is in the development of methods to enhance the mobility or reactivity of material that, by its nature, is not particularly mobile or reactive.

Three Categories of Remediation Technologies

Remediation technologies can be divided into three general categories: (1) technologies for solidification, stabilization, and containment; (2) technologies exploiting biological and chemical reactions to destroy or transform the contaminants; and (3) technologies involving separation of the contaminants from the contaminated media, mobilization of the contaminants, and extraction of the contaminants from the subsurface. Box 3-3 provides definitions of the different types of technologies in each of these three categories.

Solidification and stabilization processes are directed at decreasing the mobility and/or toxicity of contaminants by reducing contaminant solubility or volatility and medium permeability. Most such techniques have been developed for ex situ treatment of soil contaminated with heavy metals, although a few methods for in situ treatment of relatively shallow contaminated soils are in use. These processes are generally not suited for contaminants located at significant depth or for very volatile or soluble organic contaminants, although some of the methods are now being applied to a limited number of organic contaminants.

Containment methods are designed to prevent movement of contaminants away from the zone of contamination by providing a physical or hydraulic barrier. Low-permeability clay and/or geotextile caps and low-permeability slurry walls are fairly standard technology. Combinations of reactive processes with physical containment systems are a new innovation being implemented in the field. Pump-and-treat systems are also often used to hydraulically contain contaminated ground water.

Biological and chemical reaction processes use biological or chemical reactions to transform contaminants to innocuous, or at least less harmful, products. Biological processes, known generally as bioremediation, rely on microorganisms to mediate contaminant transformation reactions and degrade the compounds. Many organisms native to soils can use contaminants as sources of carbon and energy for growth. Some organisms (known as aerobes) require oxygen to thrive, while others (known as anaerobes) thrive in oxygen-free environments and use other electron acceptors, such as nitrate, iron, sulfate, and carbon dioxide. Addition of nutrients, moisture, or the appropriate electron acceptors can increase microbial activity and thus enhance the reaction rates. Pretreatment with enzymes or chemical oxidants can make complex chemicals more readily degradable. For in situ bioremediation applications, the primary challenges are largely related to creating the necessary environmental conditions in situ that will cause biodegradation of the contaminants; this includes delivery of the necessary amendments to contaminated locations. Chemical reaction processes are not used as frequently as biological processes. Few chemical reaction processes are available, and even fewer have been tested extensively. However, several technologies are now being tested, as shown in Box 3-3. Reaction processes, whether biological or chemical, are the only processes that can completely destroy organic contaminants. The

BOX 3-3
A Glossary of Remediation Technologies

Stabilization/Solidification and Containment Technologies

Asphalt batching. Encapsulates contaminated soil in an asphalt matrix. Volatile organic compounds (VOCs) generally volatilize during the process and are captured and treated in an off-gas system.

Biostabilization. An ex situ microbial process to rapidly degrade the bioavailable components (the more volatile and soluble fractions of contaminant mixtures). The process leaves behind a much less mobile and less bioavailable residue.

Enhanced sorption. A passive-reactive barrier (see definition below) that creates zones that cause contaminant sorption, either microbiologically (biosorption) or chemically (using surfactant coatings).

In situ precipitation/coprecipitation. A passive-reactive barrier (see definition below) that causes the precipitation of a solid (usually carbonate, hydroxide, or sulfide mineral) to maintain a toxic metal in an immobile form. Formation of solid phases is controlled primarily by pH, redox potential, and concentration of other ions.

In situ soil mixing. A method of achieving stabilization of contaminated soil in situ. Soil is mixed with stabilizing agents using large augers in successive drillings across a site.

Lime addition. A method that decreases permeability of soils by filling interstitial pore spaces and forming weak bonds between soil particles. Heat generated due to hydration can aid in thermal desorption of VOCs.

Passive-reactive barriers. Permeable containment barriers that intercept contaminant plumes and remove contaminants from ground water solution using chemical and/or biological reactions within the barrier.

Pozzolonic agents. Cement-like materials that form chemical bonds between soil particles and can form chemical bonds with inorganic contaminants, decrease permeability, and prevent access to contaminants. The most common pozzolonic materials are portland cement, fly ash, ground blast furnace slag, and cement kiln dust.

Slurry walls, sheet pile walls, and *grout walls.* Low-permeability barriers designed to prevent contaminant transport in situ. The success of these technologies depends on achievement of a long-lived, low-permeability barrier. Because these walls create hydraulic confinement, fluid must either be allowed to flow around them or be removed from the system and treated if necessary. Alternatively, the barrier wall must contain permeable zones in which reactions can occur.

Vitrification. Melting of contaminated soil combined with amendments as needed to form a glass matrix from the soil, either in place (in situ vitrification) or in a treatment unit. Nonvolatile metals and radioactive contaminants become part of the resulting glass block after cooling. Organic contaminants are either destroyed or volatilized by the extremely high temperatures. The method is generally expensive due to large energy requirements.

Biological Reaction Technologies

Biopile. Soil placed around or over ventilation pipes and often amended with nutrients (during emplacement, or by irrigation or batch additions). Biopiles are supplied with oxygen by vacuum-induced air flow.

Bioslurry reactor. Biological (ex situ) reactors that slurry, suspend, and typically aerate solids. Reactors can be enclosed or in the form of treatment lagoons. When volatile constituents are present, vapor capture and treatment may be necessary. This technology is usually used for sludges with high water content.

Biostabilization. See above definition (under "Stabilization, Solidification, and Containment Technologies").

Bioventing and *biosparging.* A form of engineered in situ bioremediaiton involving addition of oxygen to stimulate aerobic microbial activity. Oxygen is added by soil vapor exchange (bioventing) in the vadose zone and by air sparging (biosparging) in the saturated zone. Air flow is generally lower in bioventing and biosparging than in SVE and air sparging systems, which are designed to maximize extraction of volatile components from the subsurface.

Composting. Related to biopiles. Additional carbon, in the form of manure, sludge, plant byproducts, or wood chips, is added to increase biological activity and pore size.

Engineered in situ bioremediation. Addition of electron acceptors (usually oxygen) or donors and nutrients in situ to ground water or soil to facilitate biodegradation. Biodegradation occurs in the ground water system downgradient from the point of nutrient addition. The systems generally do not require large energy inputs.

Enhanced sorption. See above definition (under "Stabilization, Solidification, and Containment Technologies").

Fungal treatment. Addition of wood-degrading fungi, either white rot or brown rot, to a biopile or land farming application. The fungi degrade complex organic compounds by producing extracelluar enzymes.

Intrinsic bioremediation. The use of native soil microorganisms to degrade contaminants without human intervention other than careful monitoring. The method can be used both to destroy contaminants and to control the spread of contaminant plumes. It requires monitoring and modelling to document the existence and rate of biodegradation. Treatment time can be very long (decades).

Land farming. Spreading of contaminated soil over a prepared bed on the land surface in shallow lifts followed by tilling to provide aeration. Treatment time depends on contaminant and soil properties, including the rate of compound release from the solids. Tilling and nutrient addition frequencies can affect remediation rates.

continued on next page

BOX 3-3 (continued)

Organic biofilters. A form of engineered in situ bioremediation that uses a large mass of microorganisms for sorption or transformation of contaminants. Electron acceptors and nutrients are added to sustain the microorganisms.

Passive-reactive barriers. See above definition (under "Stabilization, Solidification, and Containment Technologies").

Phytoremediation. Remediation of contaminated soil in situ using vegetation.

Sparge barriers. A form of engineered in situ bioremediation in which oxygen is provided to the subsurface via air injection wells placed directly into the formation or in a permeable trench.

Chemical Reaction Technologies

Chemical oxidation. Use of strong oxidants to destroy organic contaminants. The process works best on compounds, such as olefins and substituted aromatics, that contain unsaturated carbon-carbon bonds. Several chemical combinations can be used: peroxide, peroxide and iron (Fenton's reagent), ozone, hydrogen peroxide and ozone (peroxone), and potassium permanganate.

Incineration. Oxidation of organic compounds at extremely high temperatures (ex situ). Organic compounds that are difficult to treat by other methods can be destroyed by incineration.

Substitution. Use of ex situ organic chemical reactions to convert soil contaminants into components that are less toxic or unregulated, typically by replacing a halogen with a hydrogen or functional group, such as an ether.

Thermal reduction. Use of hydrogen (ex situ) at elevated termperatures to reduce and decompose organic contaminants in soils to nontoxic compounds.

Zero-valent iron barrier. A passive-reactive barrier (see above definition) that creates very reducing conditions, resulting in hydrogen generation. Dissolved chlorinated solvents (ethenes, ethanes, and methanes) are chemically degraded at relatively rapid rates. Some metals form relatively insoluble solids at low redox potential and can be treated with this method.

Separation, Mobilization, and Extraction Technologies

Air sparging. Injection of air under pressure below the water table in unconfined aquifers. The method removes VOCs by volatilization while incidentally stimulating aerobic biodegradation processes. It is applicable in permeable and homogeneous soils.

Cosolvent flushing. Addition of a solvent to significantly increase the solubility of nonaqueous-phase liquids (NAPLs) and, in the case of heavy organic mixtures, to reduce overall NAPL viscosity and improve recovery. Cosolvents greatly increase the solubility of sorbed organic contaminants.

Dual-phase extraction. A process of simultaneously removing water and air from a common borehole by the application of a high vacuum. The process dewaters the area to be treated and subsequently removes the contaminant by volatilization.

Electrokinetics. The migration of chemicals through a soil matrix under the application of electrical and hydraulic gradients to effect contaminant removal. The process can function in both saturated and unsaturated environments.

Electroosmosis. Use of an electrical potential to cause movement of pore water through a clay aquifer formation to treatment zones. This technique has been long understood as a means to control water movement in fine-grained media and is currently being investigated at waste sites to treat contaminants in ground water.

Fracturing technology. Injection of fluid under pressure into the soil matrix to break up the soil and facilitate movement of treatment fluids. The process employs the principle that if the overburden pressure is exceeded, the soil will fracture, creating fissures. Both pneumatic and hydraulic fracturing are employed. In many cases, a prop material is injected into the fracture at the time of fracturing or before the pressure is released to keep the fracture open while filling it with a transmissive material. Fractures generally occur along weak points in the soil matrix, such as in preexisting fractures, lenses, bedding planes, discontinuities, or desiccation cracks. In some cases, the soil may be notched to promote fractures at a particular horizon or in a particular direction. Once created, fractures provide a transmissive pathway for the injection or extraction of fluids.

In situ soil mixing. Use of augers or impellers to break apart the soil structure and increase its transmissivity. The increase in transmissivity is accomplished by the disruption of the soil matrix, creating channels throughout the soil. A diluent or bulking agent can be added to further increase transmissivity.

NAPL recovery. The physical removal of separate-phase organic liquids. The simplest form is a gravity drainage system, in which NAPLs flow into downgradient collection points. For light NAPLs (LNAPLs), recovery can be enhanced by depressing the water table and increasing the hydraulic gradient. The rate and extent of product recovery are inversely related to NAPL viscosity and proportional to hydraulic gradient; recovery is greatest for lighter products such as gasoline, diesel, and jet fuel. Dense NAPLs (DNAPLs) may be pumped from a depression in a confining layer located at the interface between relatively coarse- and fine-grained media.

Pump-and-treat system. A process for removing dissolved contaminants

continued on next page

BOX 3-3 (continued)

from ground water by pumping the water to the surface and treating it. The process is effective for controlling and diminishing the size of plumes of dissolved contaminants. However, for source areas, it is effective only as a containment or control method due to the low solubility and large masses of contaminants present.

Soil flushing. An in situ process that uses chemical amendments and fluid pumping to mobilize and recover contaminants (see also cosolvent flushing and surfactant flushing).

Soil vapor extraction (SVE). The removal of volatile organic contaminants from unsaturated soils by inducing air flow and thus speeding contaminant volatilization. The treatment rate is a function of the volatility of the contaminant and the ratio between the air flow rate and contaminant mass. A secondary benefit of technology is that it stimulates aerobic biodegradation.

Soil washing. An ex situ process that first segregates the most contaminated soils and then washes them with a water-based solution. Generally, soil fines have a high concentration of contaminants, while coarse materials may be sufficiently clean that contaminant concentrations are below action levels, allowing coarse materials to be disposed of separately. Once fines are separated from coarse soils, the fines are washed with a solution that may include surfactants, acids, chelating agents, or other amendments to enhance desorption and solubilization.

Steam sparging. Addition of steam to enhance contaminant volatilization in air sparging systems. Treatment rates are significantly faster than in standard air sparging systems.

Surfactant flushing. Application of anionic and nonionic surfactants in situ to enhance the removal of organic contaminants. Using surfactant doses greater than the critical micelle concentration for the surfactant (usually greater than 0.5 to 1.0 percent in soils), organic contaminants partition into mobile micelles, allowing them to be transported with ground water at concentrations many orders of magnitude greater than would otherwise be possible. The contaminant-laden solution is collected and treated ex situ. Surfactants also decrease surface tension between the NAPL and water and can cause remobilization of NAPL. Remobilization can be used to increase recovery of LNAPLs, but it is generally not considered favorable in most DNAPL treatment schemes. Variations on surfactant flushing systems include the use of foams and stable gases.

Thermal desorption. A process (different from incineration) that removes volatile and semivolatile organic compounds from excavated soils by transfer to a gas phase. Volatilization is a function of compound volatility, surface area, and temperature. The simplest desorbers use soil shredding to expose surface area. Either hot gases are applied to the soil, usually in a rotary kiln or fluidized bed, or heat is transferred by solid-solid contact with the contaminated soil as it travels along a heated screw or conveyor. The vaporized contaminants are captured and collected or destroyed. The process may be augmented with the addition of quick lime, which generates heat as it hydrates. With compounds having medium to low volatility, thermal energy is added to enhance volatilization.

Thermally enhanced NAPL recovery. Use of steam or radio-frequency energy to supply heat and reduce NAPL viscosity. For heavy fuels or oils, the process can increase recovery by as much as an order of magnitude.

Thermally enhanced SVE. Addition of thermal energy to accelerate SVE cleanups or extend SVE application to less volatile organic mixtures such as diesel or fuel oil. Thermal energy can be supplied by steam, hot air, radio waves, microwaves, or electrical resistance.

Vacuum-assisted NAPL recovery. Application of a vacuum to reduce interstitial pressure, allowing NAPLs to move through soil more easily. Vacuum assistance can increase the rate and ultimate amount of product recovery by several fold.

In above-ground bioremediation cells (under construction), oxygen is supplied to soil via a network of embedded piping. Courtesy of Fluor Daniel GTI.

Monitoring an air sparging and soil
vapor extraction system. Courtesy
of Fluor Daniel GTI.

vast majority of the practices in this category are biological treatment methods
for hydrocarbon-contaminated sites. For all other classes of chemicals, far fewer
tested reactive treatment options are available.

Separation, mobilization, and extraction processes are designed to separate
contaminants from geologic materials in the subsurface, mobilize them into the
ground water or air in soil pores, and extract them from the subsurface. Some of
these technologies use heat, chemicals, vacuums, or electrical currents to separate
the contaminants from geologic materials and move the contaminants to a location where they can be extracted. For example, heat has a pronounced effect on
decreasing the viscosity of nonaqueous-phase liquids (NAPLs) and increasing
the vapor pressure of organic chemicals, making it easier to mobilize and extract
them. Other technologies in this category alter the physical structure of the soil
matrix by fracturing or mixing it, which facilitates the addition or extraction of
fluids for subsurface treatment. Separation, mobilization, and extraction processes
can enhance the efficiency of conventional pump-and-treat or SVE systems.

Availability of Remediation Technologies for Various Problem Contexts

Table 3-1 shows the availability of technologies in the three categories for treating four different types of contaminated media: (1) surface soils, sediments, and sludges; (2) the unsaturated zone (soil contamination below the surface but above the water table); (3) the saturated zone (contamination below the water table); and (4) subsurface source zones. As shown in Table 3-1, the availability of technologies is greatest for treatment of surface soils, sediments, and sludges. The deeper and more entrenched the contamination problem, the more limited is the menu of technology options.

Table 3-2 shows the availability of the three general categories of technologies for treating different classes of contaminants. As shown in the table and in Figure 3-1, a range of treatment alternatives has been developed for the relatively mobile and biodegradable contaminants (petroleum hydrocarbons and chlorinated solvents). The number of potential treatment technologies is much smaller for the other classes of chemicals. The remainder of this chapter discusses in detail the technology options for treating the six categories of contaminants shown in Table 3-2. These classes of contaminants are representative of contaminants typically found at hazardous waste sites.

CLEANUP OF PETROLEUM HYDROCARBONS

Sources

The presence of petroleum hydrocarbons in the subsurface is generally related to the transport, distribution, and use of fuels and oils. There are five main sources of petroleum hydrocarbon contamination: underground or above-ground storage tanks, tanker trucks, transfer terminals, pipelines, and refineries. Contamination in the subsurface typically is a result of leakage or spillage (slow, periodic, or catastrophic) or of disposal of wastes (separator sludges, waste oils, and refinery sludges and residuals). The vast majority of hydrocarbon-contaminated sites are associated with underground storage tanks. As shown in Table 1-2 in Chapter 1, there are an estimated 300,000 to 400,000 leaking underground storage tanks in the United States. In comparison, refinery and pipeline sites are fewer but typically much greater in both affected area and volume of contaminant released. According to the American Petroleum Institute, there are approximately 150 refineries in the United States (API, 1996).

Fate

Hydrocarbons are biodegradable and volatile with moderate to low solubility. Hydrocarbon contaminants in the subsurface can be found distributed among four phases: (1) sorbed to solids, (2) as an NAPL, (3) dissolved in the ground

TABLE 3-1 Technology Types Applicable to Different Contaminated Media

Context	Solidification, Stabilization, and Containment	Biological and Chemical Reaction	Separation, Mobilization, and Extraction
Surface soils, sediments, and sludges	Excavation, e Pozzolanic agents, e, i Lime/fly ash, e, i Vitrification, e, i Asphalt batching, e	Biopiles, e Composting, e Land farming, e (Fungal treatment, e) Bioslurry systems, e, (i) Incineration, e Phytoremediation, i (Biostabilization, e, i) Substitution, e Thermal reduction, e	Solvent extraction, e Thermal desorption, e Soil washing, e (Electrokinetic systems, e, i)
Unsaturated zone	Deep soil mixing Excavation (Polymer walls) Grout walls Slurry walls Sheet pile walls	Bioventing	SVE Thermally enhanced SVE (Soil flushing with surfactants or cosolvents) (Electrokinetic systems)
Saturated zone	Excavation (Polymer walls) Grout walls	Engineered in situ bioremediation Intrinsic bioremediation Biosparging	Pump-and-treat systems Sparging: air and steam (Electrokinetic systems)

	Slurry walls Sheet pile walls (Passive-reactive barriers using enhanced sorption)	(Chemical oxidation/reduction) Passive-reactive barriers -using iron reactions -using organic or microbiological reactions (-using enhanced sorption) -using nutrient additions	Dual-phase recovery (Soil flushing with surfactants or cosolvents)
High-concentration source areas in the saturated zone	Pump-and-treat systems Grout walls Slurry walls (Polymer walls) Sheet pile walls	Biosparging Bioventing (Chemical oxidation/reduction) Engineered in situ bioremediation Intrinsic bioremediation	NAPL recovery Dual-phase extraction (Soil flushing with surfactants or cosolvents) Sparging: air and steam

NOTE: Applications that are not commercially available are shown in parentheses. For surface soils, sediments, and sludges, "e" signifies an ex situ treatment method, and "i" signifies an in situ method. Technologies for surface soils, sediments, and sludges are predominantly ex situ processes, although there are some in situ alternatives. Because these technologies can potentially be applied either in situ (i) or ex situ (e), the modes of application are noted in the table.

TABLE 3-2 Treatment Technology Options for Different Classes of Contaminants

	Petroleum Hydrocarbons	Chlorinated Solvents	Polycyclic Aromatic Hydrocarbons and Semivolatile Organic Compounds	PCBs	Inorganic Chemicals	Pesticides and Explosives
Solidification, Stabilization, and Containment						
Asphalt batching	X	na	X		X	
Biostabilization	X	na	?	?	na	
Excavation (stabilization)	X	X	X	X	X	X
Grout walls			X	X	X	
Lime addition	X(h)	na			X	
Passive barriers using sorption or precipitation	?	?	?		?	
Polymer walls		?			?	
Pozzolanic agents	X(h)	na	?	?	X	
Pump-and-treat systems	X	X	X	na	X	X
Sheet pile walls		X	X	X		X
Slurry walls		X	X	X		X
Vitrification	na	na	na	?	X	
Chemical and Biological Reaction						
Biopiles	X	na	X	?	na	X
Bioslurry systems	X	na	X	?	na	X
Biosparging	X(l)	?	?		na	na
Bioventing	X(l)	na	?		na	na
Chemical oxidation	?	?	?		na	?
Chemical reduction		?			X	X
Engineered bioremediation (in situ)	X	?	?		?	?
Incineration	X	X	X	X	na	X
Intrinsic bioremediation	X	X	?	?	na	?

Land farming	X	na	X	?	na	X
Passive-reactive barriers						
-using iron	na	X	na	na	X	
-using organic/microbiological reactions	X	X	?	?	?	
-using enhanced sorption	na	?	?		?	
-using passive/active nutrient additions	X	X	?		na	
Phytoremediation	X	X	?	?	na	?
Substitution	?	?	?	X	X	?
Thermal destruction/reduction	X	X	?	na	na	
Separation, Mobilization, and Extraction						
Dual-phase extraction	X(l)	X	na	na	na	
Electrokinetic systems	na	?	na	na	?	
Soil washing	X	X	X	?	X	
Soil flushing						?
-acid, base, or chelating agent	?	?			?	
-steam	?	?	?		na	
-foam	?	?	?		na	
-surfactant/cosolvent	?	?	?	?	na	?
Sparging: air/steam	X(l)	X	?	na	na	
Recycling/re-refining	X	na	na	na	X	
Thermal desorption	X	X	X	X	na	
Thermally enhanced SVE	X(h)	X	?	X	X	X
Solvent extraction	X	na	na	na	na	
SVE	X(l)	X	na	na	na	

NOTES:

X(h): applicable primarily for heavy fuels or high-molecular-weight solvents

X(l): applicable to light hydrocarbons only

na: technology not applicable to this class of contaminants

?: application not commercially available or exists in an experimental stage

blank: lack of information for qualitative comparison

water, and (4) as a vapor in unsaturated soil. Because hydrocarbon mixtures have low solubility, most of the hydrocarbon mass is typically in the sorbed or NAPL phase. For example, a typical phase distribution of gasoline in sand is 30 to 50 percent as NAPL, 40 to 50 percent sorbed, 2 to 5 percent dissolved, and less than 0.5 percent in the vapor phase (Brown et al., 1987b).

When hydrocarbon liquids are released to soil, they migrate downward until they are retained as a residual in soil pores. The amount of hydrocarbon liquid retained is a function of the fluid viscosity and the soil texture, which vary widely. More viscous (heavier) hydrocarbon mixtures and fine soil textures generally result in greater hydrocarbon retention within the soil. The residual hydrocarbon concentration in soil ranges from approximately 10,000 to 15,000 mg/kg for gasoline in fine sand to 60,000 to 80,000 mg/kg for no. 6 fuel oil in fine sand (Lyman et al., 1992). Because hydrocarbon mixtures are less dense than water, they typically accumulate in a layer on the water table when sufficient hydrocarbon has spilled or leaked to saturate the soil, allowing free-phase liquid to migrate to the water table.

Three mechanisms serve to attenuate petroleum hydrocarbon liquids in the subsurface: (1) biodegradation, (2) volatile transport and exhaust to the soil surface, and (3) dissolution. Because the effective solubilities and vapor pressures of the various components of hydrocarbon mixtures are low, removal of these source materials by solubilization or volatilization is slow. Biodegradation is usually the more significant mechanism for attenuation of hydrocarbons except near the soil surface, where volatilization may play a more significant role. Biodegradation is relatively slow for sorbed hydrocarbons, but biodegradation of dissolved hydrocarbons is relatively rapid.

Biodegradation is carried out by ubiquitous native soil microorganisms (Claus and Walker, 1964; Alexander, 1994; Chapelle, 1993). The number of hydrocarbon-degrading microorganisms is much greater in hydrocarbon-contaminated sediments than in uncontaminated zones (Aelion and Bradley, 1991). The rate of biodegradation and the metabolic products produced are controlled primarily by the types of hydrocarbons present and the availability of electron acceptors and nutrients needed by the microorganisms to conduct the reactions (National Research Council, 1993). Aerobic biodegradation is more rapid than anaerobic biodegradation, but oxygen is generally limited in the immediate vicinity of subsurface hydrocarbons because of the low solubility of oxygen and the high oxygen demand created by the hydrocarbon-degrading organisms. In the absence of sufficient oxygen, microorganisms can use alternative electron acceptors such as nitrate, iron, sulfate, and carbon dioxide to biodegrade hydrocarbons (Chapelle, 1993; Hutchins and Wilson, 1994; Barbaro et al., 1992; Wilson et al., 1994).

Dissolution, volatilization, and biodegradation do not rapidly or completely remove hydrocarbon mass from the subsurface. However, these processes together cause weathering of the petroleum product. As a product weathers, the

TABLE 3-3 Treatability of Petroleum Hydrocarbons

Hydrocarbon	Volatile Fraction (Percent)	Degradability	Treatability
BTEX solvents	100	High	Very high
Gasoline	>95	High	High
Jet fuel	75	High	High
Diesel/kerosene	35	High	Moderate-high
No. 2 fuel oil	20	Moderate	Moderate
No. 4 fuel oil	10-20	Low-moderate	Low-moderate
Lube oil	10-20	Low-moderate	Low-moderate
Waste oils	<10	Low	Low
Crude oils	<10	Low	Low

SOURCE: Brown and Norris, 1986.

more mobile (volatile and soluble) and degradable fractions are removed. The remaining residue is more viscous and less soluble than the original contaminant mixture, reducing the risk of continued contamination of soil and ground water.

Remediation Technology Options

Because some components of petroleum hydrocarbons are relatively mobile and biodegradable compared to other types of contaminants, a large number of technologies are applicable to hydrocarbon remediation. Applicable technologies include NAPL recovery, dual-phase extraction, in situ bioremediation, biopiles, land farming, SVE, bioventing, biosparging, soil washing, and soil flushing. The processes that can be applied to various sources of hydrocarbon contaminants vary considerably and are a function of the type of hydrocarbon product. In general, which remediation technologies will be applicable to hydrocarbon contamination is a function of the mobility and reactivity of the hydrocarbon. Mobility and reactivity, in turn, are functions of the properties and quantities of the particular hydrocarbon and the hydrogeologic setting in which it is found. In general, lighter hydrocarbons are more volatile and degradable and thus more readily treatable than other types of hydrocarbons. Table 3-3 shows the treatability of various petroleum hydrocarbon products (Brown et al., 1987a).

Separation Techniques

Soil Vapor Extraction.   SVE removes petroleum hydrocarbons by two mechanisms: volatilization and biodegradation (P. Johnson et al., 1990; R. L. Johnson et al., 1992). Volatilization occurs when the air stream contacts residual hydrocarbons or films of water containing dissolved hydrocarbons in soil. Biodegradation occurs because the induced air flow supplies oxygen for aerobic biodegradation. All petroleum hydrocarbon fuels are essentially biodegradable (Chapelle, 1993). However, the volatile fraction, and therefore the rate of treatment of hydrocarbons, varies. Hydrocarbon fuels with a high volatile fraction will be removed most rapidly using SVE; those with a low volatile fraction will be less responsive. Volatility ranges from more than 90 percent for gasoline to less than 10 percent for crude oil (see Table 3-3). Based on approximate volatilities, SVE is a primary technology for the remediation of gasoline, jet fuel, and mineral spirits. SVE can be used to treat the other, less volatile hydrocarbon mixtures as part of a biodegradation strategy, a process often termed bioventing.

SVE is commonly limited by the permeability of the soil and by the degree of saturation. SVE will not work well in low-permeability soils such as silts and clays or in highly saturated areas, such as the capillary fringe or below the water table.

SVE is a widely used commercial technology for the treatment of petroleum hydrocarbon releases; as of 1995, it was in use or had been used at 139 Superfund sites and nearly 9,000 underground storage tank sites (see Figures 1-7 and 1-8 in Chapter 1). It has moderate to high success in achieving specific regulatory goals. Generally, SVE is most successful for treating more volatile hydrocarbon products and more permeable soils.

Soil Washing and Soil Flushing.   Significant quantities of petroleum hydrocarbons can be retained in soils as a residual, discontinuous NAPL phase. One approach to removing residual petroleum products is to use surfactants or cosolvents. Surfactants and cosolvents can desorb hydrocarbons from soils and can decrease the interfacial tension of the NAPL, forcing it from the soil matrix and allowing it to coalesce into a recoverable, continuous NAPL phase (Gotlieb et al., 1993).

There are two basic types of surfactant and cosolvent applications. One is soil washing, which is a process for removing hydrocarbons from excavated soils. The other is soil flushing, which is the in situ application of surfactants or cosolvents to contaminated soils. Both of these processes have significant variations. With both types of applications, site-specific blends of additives are generally used. A significant portion of the cost is associated with unrecovered additives and disposal of generated fluids.

In soil washing, the petroleum-contaminated soil is excavated, slurried, and processed. In some soil washing systems, the soil slurry is processed by soil sizing to concentrate the hydrocarbons in the finer soil fractions. The surfactant or cosolvent is then added to the fine soil slurry fraction, minimizing the amount of additives required. Other systems add the remedial agent directly to the soil and then agitate the slurry. The soil, water, and NAPL phases are then separated. Some soil sizing may be used to enhance the separation (coarser fractions are easier to dewater). Soil washing is used commercially to treat petroleum-contaminated soils (Delta Omega Technologies, 1994). It is not as commonly used to treat lighter, more volatile products such as gasoline, jet fuel, or mineral spirits, because removing these products by volatilization is more cost effective. While soil washing is used to treat a wide range of soil types, it has limited applicability to soils with high clay content due to problems in separating the fine clay particles.

There are three principal variations to in situ soil flushing: (1) enhanced solubilization, (2) emulsification, and (3) displacement. In the first approach, chemical additives (such as surfactants and cosolvents) are used to enhance the aqueous solubility of contaminants in order to more efficiently dissolve or desorb the petroleum hydrocarbons (or other organic contaminants). In the second approach, higher concentrations of these additives are used to emulsify the NAPLs, either as microemulsions or middle-phase emulsions, and flush them out more effectively. The contaminant molecules dissolve into mobile micelles of the additive, which are entrained in the water. In the third approach, additives that decrease NAPL-water interfacial tensions to very low values (less than 1 dyne/cm) are used to mobilize the trapped ganglia and displace the resulting bank of free-phase liquid. The first approach involves miscible displacement (i.e., resident and introduced fluids mix completely), while the other two methods involve immiscible displacement (two immiscible fluids—oil and water—are displaced). Combinations of various surfactants and cosolvents can be used to achieve solubilization, emulsification, or displacement. In situ flushing with steam has also been attempted. All of these technologies have been tested and used for enhanced recovery in oil fields, but use for site remediation purposes has been limited to several pilot-scale and a few commercial-scale tests (Grubb and Sitar, 1994; EPA, 1995a).

Thermal Desorption. Thermal desorption is a commonly used technology for treating excavated petroleum-contaminated soils. It is based on the principle that volatility increases with increasing temperature. What distinguishes thermal desorption from incineration is that the soil does not contact a flame. The petroleum product is volatilized off the soil, and the resulting vapor stream is captured and treated. There are two variations: low temperature and high temperature. Low temperature thermal desorption uses temperatures of less than 200°C (400°F). It is used to treat more volatile products such as gasoline, jet fuel, mineral spirits, and sometimes diesel. High temperature thermal desorption uses temperatures of 320 to 430°C (600 to 800°F). It is used to treat soil contaminated with diesel and fuel oil. Neither process is effective with very heavy products such as no. 6 crude oils. With thermal desorption, the heat is applied either through hot air or through radiant or convectional heating. With hot air systems, a fuel is combusted, and the combustion gases are fed into the desorption unit.

Thermal desorption is used for a wide range of products, but use is most common for motor fuels (gasoline, diesel, and jet fuel). Units range in size from those that can process 5 to 10 tons per hour to those that can handle more than 40 tons per hour. The process is generally able to achieve regulatory standards that allow the soil to be reused or disposed of on site.

NAPL Recovery.   NAPL recovery is the removal of separate-phase liquid hydrocarbons (at amounts greater than residual saturation) from the soil matrix. NAPL recovery is best accomplished with low viscosity hydrocarbon products. Viscous products such as no. 6 fuel oil or crude oils do not flow readily through soils and are not typically recovered. Most NAPL recovery systems also produce water. Because gravity drives the collection of NAPLs, the water table is often depressed to increase the flow of NAPL into the collection point.

The variations in NAPL recovery are a function of how and where the NAPL and water phases are separated. In permeable formations and with low viscosity products, the separation process is accomplished in situ using dual pumps (product and water), automatic bailers, or oil skimmers. In low-permeability formations, a total fluid extraction system is used to recover both product and water. These are then separated on the surface with conventional technologies for separating oil and water. A recent innovation in NAPL recovery is vacuum-assisted NAPL recovery (Kittel et al., 1995). In this process, a vacuum is applied to the recovery well to promote the flow of NAPL into the collection point. The vacuum application minimizes the amount of water that is collected by creating a driving force that is an alternative to depressing the water table. The vacuum may be applied to the well bore directly or through an inner tube (drop tube). The use of a drop tube is sometimes referred to as "bioslurping" (Kittel et al., 1995).

NAPL recovery is a standard remediation technology employed at almost any site having recoverable NAPL. Most applications use water table depression as the driving force for NAPL recovery. Vacuum assisted recovery is being increasingly used because it minimizes the amount of water that needs to be treated and disposed. NAPL recovery is generally able to remove NAPLs to the point where all that remains is a thin film of oil noticeable only by its iridescent sheen on the water. It is ineffective for NAPLs present as residual saturation in soil.

Thermally Enhanced Product Recovery.   Highly viscous petroleum products such as no. 4 and no. 6 fuel oils or crude oils do not flow readily through geologic formations and are therefore not easily recovered with conventional NAPL recovery techniques. A means of promoting their recovery directly from soils is to use thermally enhanced product recovery. Viscosity is a function of temperature: the higher the temperature, the lower the viscosity. Typically, subsurface temperatures need to be in the range of 66 to 93°C (150 to 200°F) for the technology to be effective. The subsurface temperature may be raised using hot air, steam, electrical heating, or radio frequency heating. Hot air has limited application because of its low thermal capacity. The application of heat has been demonstrated to increase recovery of heavy oil products by an order of magnitude.

Thermally enhanced product recovery is a commercial technology but has limited utility (EPA, 1995b). Generally, it is used for heavy products and where steam is readily available, such as at sites with existing boilers. The cost of a transportable boiler makes this technology too expensive for routine operations.

Thermally enhanced product recovery is generally able to remove NAPLs from wells down to about a tenth of a meter (several inches). It can remove some, but not all, NAPLs present as residual saturation in soil.

Dual-Phase Extraction. Dual-phase extraction is the simultaneous removal of vapors and water from a common borehole by the application of a high vacuum. The purpose of the technology is to treat soil contamination below the water table so that the volatile components may be removed. The technology combines dewatering and venting. It is generally applied to lower permeability formations to minimize the amount of water that needs to be recovered or treated.

There are two variations of this technology. The first uses an internal drop tube to apply the vacuum to the bottom of the borehole. The drop tube removes the water in the well; once the well is dewatered, it will also remove vapors. The second variation uses conventional down-hole water pumps and applies a vacuum to the borehole. The applied vacuum aids in water removal and promotes volatilization in the dewatered soil.

Dual-phase extraction is best applied to hydrocarbon mixtures (such as gasoline, jet fuel, and mineral spirits) that have highly volatile components. It also

Sparging point being checked during regular site inspection. Courtesy of Fluor Daniel GTI.

may be used in conjunction with bioremediation to treat less volatile but degradable hydrocarbons. Use of this technology is increasing due to reports that it can be applied at low-permeability or heterogeneous sites for which few other remediation options exist (Brown and Falotico, 1994).

Air Sparging.   Air sparging is the injection of air directly into the saturated zone (Brown, 1992). The injected air treats adsorbed and dissolved hydrocarbons through volatilization and/or biodegradation. The success of air sparging depends on the distribution of air through the saturated zone and the degree of mixing of the ground water. Air sparging works best in homogeneous, moderately permeable media such as fine to medium sands.

Air sparging is used to treat both volatile and nonvolatile hydrocarbon mixtures. With volatile mixtures such as gasoline and jet fuel, air sparging operates as both an extraction (volatilization) and a transformation (biodegradation) process. With less volatile mixtures, it is primarily a means of supplying oxygen to enhance biodegradation. For volatile hydrocarbons, air sparging systems are generally applied with an SVE system to capture any released hydrocarbons.

Air sparging is more effective for treating dissolved hydrocarbon plumes than for treating source areas (Bass and Brown, 1996). Air sparging has achieved regulatory goals with little rebound in contaminant levels for plumes of dissolved hydrocarbons at numerous field sites. Because of its effectiveness in treating dissolved hydrocarbons, air sparging can be used as a barrier system, in which a line of sparge wells is placed across a plume to intercept and remove dissolved constituents. The ability of air sparging to treat dissolved contaminants has made it an alternative to conventional pump-and-treat systems.

When air sparging systems are used to treat contaminant source areas, there is a higher probability of rebound of contaminant concentrations after treatment, especially when NAPLs are present. The use of air sparging to treat source areas requires close well spacing and moderate to high air flows.

A related technology is "biosparging," which uses low air flows to minimize the amount of volatilization, so that any volatilized hydrocarbons are biodegraded in the vadose zone before being discharged to the atmosphere. This technique eliminates the need for an SVE system to accompany the sparging system.

Air sparging is a commonly used technology, especially for gasoline contaminated sites. It is also used, although less commonly, at sites contaminated with diesel and jet fuel.

Biological Reaction Techniques

As Table 3-3 shows, bioremediation techniques (including biopiles, land farming, bioventing, biosparging, sparge barriers, and intrinsic bioremediation, as well as other bioremediation systems) are widely applicable for the control and remediation of petroleum hydrocarbons. Many of the current bioremediation technologies on the market were initially developed to treat petroleum hydrocarbons.

The rate at which petroleum products biodegrade varies. Generally, the heavier the product, the slower the rate of biodegradation. Natural hydrocarbon biodegradation rates can be enhanced when the substance that most limits microbial growth is supplied to the contaminated zone. This premise provides the basis for most bioremediation processes. Because the hydrocarbon contaminants supply carbon for growth, in most cases the growth-limiting factor is the electron acceptor. In unsaturated soil, oxygen can be supplied by increasing air circulation. SVE, bioventing (see Box 3-4), and biopiles are three technologies designed to increase air circulation. The U.S. Air Force has applied bioventing systems at sites across the country, with consistent hydrocarbon degradation rates of 2.4 to 27 mg hydrocarbon per kg soil per day at soil temperatures between 4 and 25°C (39 and 77°F) (Ong et al., 1994). In the presence of sufficient oxygen, elements such as nitrogen, phosphorus, and potassium in nutrient-poor soils can limit microbial growth and biodegradation (Aelion and Bradley, 1991; Armstrong et al., 1991; Allen-King et al., 1994a,b), and addition of these limiting nutrients can also enhance biodegradation rates (Allen-King et al., 1994b).

Oxygen (O_2) can also be added to the saturated zone using one of several methods (Brown et al., 1990). The first bioremediation systems used aerated water, but these systems were limited by the relatively low solubility of O_2 in water (8-12 mg/liter) relative to air. Typically, about 2 to 3 g of O_2 are required per g of hydrocarbon for complete mineralization; only about 3 mg/liter of total dissolved hydrocarbon can be mineralized in water saturated with respect to atmospheric O_2. The next generation of bioremediation technology used hydrogen peroxide (H_2O_2) to stimulate saturated-zone biodegradation (Brown et al., 1993). Air sparging is currently the most common method for supplying O_2 for enhanced biodegradation (Brown and Jasiulewicz, 1992). Solid O_2-releasing sources can also be used to promote biodegradation by adding O_2 to the ground water in situ as it flows through a permeable barrier (Bianchi-Mosquera et al., 1994).

With petroleum hydrocarbons, intrinsic remediation is a significant process. Intrinsic remediation is the reliance on natural processes, including volatilization, sorption, dilution, reactions with naturally occurring chemicals, and, most commonly, biodegradation, to decrease contaminant concentrations without human intervention other than careful monitoring. Intrinsic bioremediation (the type of intrinsic remediation in which biological processes predominate) has been well documented to occur in plumes of dissolved petroleum hydrocarbon contaminants. As documented in a survey of sites in California, petroleum hydrocarbon plumes reach an equilibrium point, often within 60 to 90 m downgradient of the source, beyond which ground water contamination generally does not pass (Rice et al., 1995). The location of the equilibrium point depends on the size of the source area, the ground water flow rate, and other environmental conditions. Equilibrium is reached though a combination of anaerobic and aerobic degradation

BOX 3-4
History of Development of Bioventing

The development of bioventing illustrates the evolution of a technology driven by market need. The market for hydrocarbon treatment technology has been significant because of the widespread use and environmental release of hydrocarbon fuels by government, industry, and the public and the enactment of legislation requiring contaminated site cleanup. While technology existed for the treatment of hydrocarbon contamination, the cost and complexity of treatment often precluded widespread use of these technologies other than in areas of heightened exposure, such as at retail gasoline stations in urban and suburban areas. The Department of Defense (DOD), in particular the Air Force, has a large number of fuel handling and storage areas, many of which have associated environmental problems. Many of these sites are in remote locations where the installation and operation of treatment systems is difficult and costly. Thus, there was a need for a simple but effective technology that could address hydrocarbon contamination in remote areas. This need for inexpensive but effective treatment technology was the market driver for the development of bioventing.

The basis for bioventing lies in two technologies: SVE (see Box 3-2) and bioremediation. Early in its application, SVE was considered a very cost-effective technology as long as the recovered vapors could be discharged directly to the atmosphere without treatment. However, concerns about air quality necessitated the use of vapor treatment, which significantly raised costs. Parallel to the development of SVE, developers of in situ bioremediation recognized that oxygen supply was a key to stimulating the biodegradation of hydrocarbons (see Box 3-1). However, the oxygenation systems used for in situ bioremediation were either ineffective or costly. Early in its development, the potential of SVE to supply oxygen and stimulate biodegradation was recognized (Thorton and Wooten, 1982; Texas Research Institute, 1982; Ely and Heffner, 1991). Despite these parallel developments, SVE and bioremediation remained separately applied technologies. Keeping SVE and bioremediation apart were concerns that bacteria would be unable to effectively scavenge oxygen from an SVE flow stream and that typical SVE systems generated considerable vapors, which often required costly vapor collection and treatment. Thus, practitioners believed that SVE would be an ineffective and costly form of bioremediation.

Several factors changed the separate application and development of SVE and bioremediation and led to the emergence of bioventing. First was the recognition that bacteria were able to effectively use oxygen from an SVE system. Early work at Hill Air Force Base demonstrated that, even at high SVE flow rates, oxygen levels were significantly depleted due to hydrocarbon biodegradation activity (Hinchee et al., 1989), demonstrating the ability of bacteria to scavenge oxygen from an air stream. Second was the recognition that vapor levels produced during SVE operation were a function of the rate of air flow and could be controlled. (Hoag and Bruel, 1988; Johnson and Ettinger, 1994). Third was the finding that many fuels at DOD sites had diminished concentrations of volatile hydrocarbon components compared to gasoline, making the control of vapor emissions even less problematic.

Based on these findings, researchers postulated that an air-based bioremediation system could be developed; the rate of volatilization could be balanced with the rate of biodegradation so that there would be no appreciable volatile discharge. Early test work on Air Force sites demonstrated that such a balanced system could be designed and operated (Miller et al., 1990; Hinchee and Ong, 1992). This early work was expanded into the Air Force's bioventing initiative, as a result of which more than 150 bioventing projects have been installed to date.

Bioventing has now evolved from an adapted form of SVE to a separate, low-cost technology. Early forms of bioventing used a vacuum-based withdrawal system augmented with nutrient addition. With the demonstration that vapor levels could be readily controlled by adjusting the air flow rate, bioventing systems switched to lower cost air injection systems. Test work demonstrated that nutrient addition was not usually necessary because oxygen is the factor limiting microbial growth, making bioventing a simple air injection system. Finally, the understanding of how hydrocarbon-utilizing bacteria scavenge oxygen has led to the development of an effective but low-cost monitoring method: in situ respirometry. With in situ respirometry, the rate of oxygen uptake and/or carbon dioxide production is used as an indicator of biodegradation activity. When the respiration rate approaches background levels (i.e., the rate determined in a nearby uncontaminated location), remediation is considered complete. This method eliminates the need for expensive soil sampling.

processes. As noted above, O_2 is limited in the immediate vicinity of subsurface hydrocarbons. In the absence of sufficient O_2, organisms will use alternate electron acceptors. Alternate electron acceptors become important when the dissolved O_2 level drops below approximately 2 mg/liter (Salanitro, 1993).

As pictured in Figure 3-2, plumes of dissolved hydrocarbons typically have an anaerobic core area surrounded by an aerobic zone (Norris and Matthews, 1994). In the anaerobic core, hydrocarbons may be degraded by denitrification, iron reduction, sulfate reduction, and methanogenesis (see National Research Council, 1993). In the aerobic zone, they are oxidized by O_2.

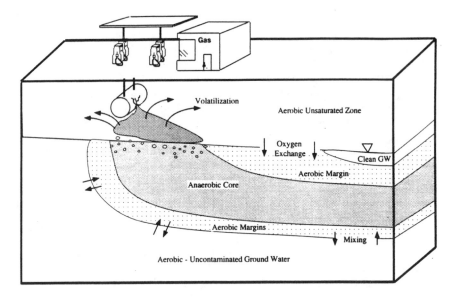

FIGURE 3-2 Plumes of petroleum hydrocarbons in ground water typically have an anaerobic (oxygen-free) core area surrounded by an aerobic (oxygen-containing) margin. Anaerobic microorganisms degrade contaminants in the core, while aerobes degrade them in the margin. SOURCE: Reprinted, with permission, from Norris and Matthews, 1994. © 1994 by Lewis Publishers.

Research Needs

While an abundance of technologies is available for cleaning up sites contaminated with petroleum hydrocarbons, some problem areas still need resolution. The main needs are technologies for treating heavy hydrocarbon mixtures and hydrocarbons in low-permeability or highly heterogeneous formations. Heavy hydrocarbons have very low solubilities, sorb strongly, and resist degradation, rendering existing technologies relatively ineffective. At the same time, the impact of heavier, less soluble hydrocarbons on ground water quality needs careful study, because lack of mobility and bioavailability may limit adverse effects. Existing technologies are most limited in cleaning up low-permeability or heterogeneous geologic media because of the reduced circulation of fluids (air, water, NAPLs) in these media; technologies are needed to improve the ability to move fluids through such media. In addition, there is a continual need for investigation of ways to optimize existing processes for the treatment of all types of petroleum hydrocarbons and for the development of more cost-effective processes for hydrocarbon treatment.

CLEANUP OF CHLORINATED SOLVENTS

Sources

Chlorinated solvent use has been ubiquitous in society since these compounds were widely introduced after World War II, although in recent years use has declined somewhat due to more stringent environmental regulations. Global use of the chlorinated solvents trichloroethylene (TCE), perchloroethylene (PCE), and 1,1,1-trichloroethane (1,1,1-TCA) in 1994 totaled 900,000 metric tons, with U.S. use accounting for 40 percent of the total (Leder and Yoshida, 1995). Users vary from large manufacturing facilities, to local businesses such as garages, photographic shops, and neighborhood dry cleaners, to homeowners. Chlorinated solvents can dissolve oily materials, have low flammability, and are fairly stable, both chemically and biologically. They are commonly used in industry as chemical carriers and solvents, paint removers, and cleaning solvents. Some of the common cleaning applications of these materials are metal degreasing, circuit board cleaning, metal parts cleaning, and dry cleaning. Chlorinated solvents are also used as intermediates in chemical manufacturing and as carrier solvents in the application of pesticides and herbicides. They have also been employed as fumigants. For a period of time, because of their solvent properties and density, TCA, TCE, and PCE were also used as household drain cleaners (Pankow and Cherry, 1996).

Because of their widespread use in industry, commercial establishments, agriculture, and homes, chlorinated solvents are among the most common ground water contaminants. Nine of the 20 most common chemicals found in ground water at Superfund sites are chlorinated solvents. TCE is the contaminant most commonly detected in ground water at Superfund sites, and PCE is third most common (National Research Council, 1994).

Fate

Chlorinated solvents may be released to the environment through the use, loss, or disposal of the neat liquids or through the use or disposal of wash and rinse waters containing residual solvents. In the latter case, the site will be affected primarily by dissolved-phase contaminants with concentrations as high as tens to hundreds of parts per million.

The movement and dispersion of chlorinated solvents in the subsurface vary depending on whether the solvents were released as a neat liquid or in dissolved form. If released in dissolved form, chlorinated solvent migration is governed largely by hydrogeological processes. The presence of solubilizing agents such as soaps (from wash waters) that counteract natural soil sorption-retardation mechanisms may facilitate the migration of the dissolved solvents. If the chlorinated solvent was released as a neat liquid, the liquid solvent will migrate downward through the soil column under the force of gravity. A portion of the solvent will be retained in the soil pores, but if sufficient solvent is present, the solvent will saturate the available soil pore space and continue moving downward until it encounters a physical barrier or the water table. When the solvent encounters the water table, it will spread out along the water table until enough mass accumulates to overcome capillary forces (Schwille, 1988). At this point, the chlorinated solvent will penetrate the surface of the water table due to the much greater density of the chlorinated solvent relative to water and travel downward by gravity until the mass of moving liquid is diminished by sorption or until it encounters an aquitard. If there is sufficient liquid mass, the solvent can accumulate along the aquitard as a DNAPL (Cohen and Mercer, 1993). (For an illustration of chlorinated solvent transport, see Figure 1-5 in Chapter 1.)

Contamination due to the release of large quantities of chlorinated solvent can comprise several distinct problems (see Figure 1-5), including gas-phase solvent in the vadose zone, sorbed solvent and residual DNAPL both above and below the water table, and dissolved-phase contamination that can occur in both shallow and deep sections of the aquifer. The amount of solvent retained by the soils can range from 3 to 30 liter/m^3 in unsaturated soils and from 5 to 50 liter/m^3 in saturated soils (Mercer and Cohen, 1990). Generally, more solvent will be retained in finer soils. Retained DNAPL can occupy as much as 5 to 25 percent of the available pore space in sandy soils (Mercer and Cohen, 1990).

In general, the difficulty of treating chlorinated solvent contamination problems is, in increasing order and without regard to the geologic matrix, as follows:

1. residual-phase solvents in the unsaturated zone,
2. dissolved-phase solvents,
3. residual-phase solvents in the saturated zone, and
4. solvents present as pools of DNAPLs.

Residual-phase chlorinated solvents in the vadose zone are the easiest to treat because of the high vapor pressure of most of these solvents and because moving air through soils is easier than moving water through the saturated zone. Dissolved-phase chlorinated solvents can be treated if there are no appreciable residual-phase solvents present—that is, if the dissolved plume is the result of the discharge of wash waters or low-level use of solvents. Residual-phase solvents in the saturated zone are treatable, but they must be located. Delineating the affected area can be the most difficult part of remediation. If the saturated zone residual-phase solvents are present in clays or fractured rock, treatment is difficult because of limited access. Solvents present as DNAPLs are the most difficult to treat because they are difficult to locate (National Research Council, 1994). There are no reliable techniques for detecting the presence of DNAPLs, and their detection is often fortuitous, or their presence may simply be inferred from contaminant concentration data in the ground water. In addition, when DNAPL sources are

TABLE 3-4 Solubilities and Vapor Pressures of Chlorinated Solvents

Compound	Solubility (mg/liter)	Vapor Pressure (mm Hg)
Methylene chloride	20,000	349
Chloroform	8,200	160
Carbon tetrachloride	800	900
1,1-Dichloroethylene	400	495
trans-1,2-Dichloroethylene	600	265
1,1-Dichloroethane	5,500	182
1,2-Dichloroethane	8,700	64.0
Trichloroethylene	1,100	57.8
Tetrachloroethylene	150	14.0
1,1,1-Trichloroethane	1,360	100

SOURCE: Cohen and Mercer, 1993.

located, they are often difficult to access because they are usually at the bottom of the aquifer. A complication that frequently results with the loss of significant quantities of neat solvents is their penetration of fractures in clays or rock (Pankow and Cherry, 1996). Flow through fractures can be rapid and can occur in both saturated and unsaturated environments. Retention of the chlorinated solvents in the fractures makes remediation a difficult process.

Once chlorinated solvents have penetrated into the subsurface, their fate is quite complex. While chlorinated solvents are stable, undergoing neither rapid chemical nor biological transformations, they are nevertheless subject to several processes (including volatilization, biodegradation, and chemical transformation) that can cause their concentrations to slowly decrease. These processes can be exploited in remediation.

Chlorinated solvents are relatively soluble and highly volatile. Thus, dissolution, dispersion, and volatilization are significant transport mechanisms. Table 3-4 provides the solubilities and vapor pressures for a number of chlorinated solvents. The aqueous solubilities are several orders of magnitude higher than drinking water standards, and thus dilution by hydrodynamic dispersion of chlorinated solvents is not a viable mechanism for managing sites contaminated with these compounds.

Recent research has demonstrated that chlorinated solvents biodegrade under certain conditions. However, there is little information on in situ rates or how to manipulate the rate of degradation. For less-chlorinated solvents (e.g., those having fewer than about two chlorine atoms per molecule), aerobic degradation can occur if sufficient O_2 is present (National Research Council, 1993). Aerobic degradation of more highly chlorinated solvents (e.g., TCE) can occur by cometabolic pathways, wherein bacteria live off a second substrate (carbon and energy source) but fortuitously degrade the chlorinated solvent (National Research Council, 1993). This may occur when chlorinated solvents exist as co-contaminants with the petroleum fuel components benzene, toluene, ethylbenzene, and xylene, provided that O_2 is not depleted, because toluene is an effective cometabolite (Chapelle, 1993). Under aerobic conditions, chlorinated solvents such as vinyl chloride, dichloroethylene (DCE), and TCE may be transformed to harmless byproducts, as was observed in pilot field tests at Moffett Field, California. In these tests, methanotrophic bacteria cometabolized chlorinated aliphatic solvents in the presence of methane and O_2 (Semprini et al., 1990).

Chlorinated solvents also may be transformed under anaerobic conditions. In this case, the chlorinated compounds undergo a process of reductive dechlorination, in which the solvents are transformed to less chlorinated compounds. For example, anaerobic microorganisms can convert TCE to DCE and DCE to vinyl chloride (Chapelle, 1993). Recent work has shown that anaerobes can in turn reduce vinyl chloride to ethene, which is in turn converted to methane, carbon dioxide, and hydrogen chloride (Chapelle, 1993). Current research is characterizing degradation of chlorinated solvents by bacterial communities that use a variety of electron acceptors, including nitrate, iron, and sulfur, as well as by methane-producing bacteria (methanogens). Most of these organisms thrive more readily on low levels of dissolved contaminants than on NAPL or sorbed-phase contaminants. High concentrations of chlorinated solvents are toxic to microorganisms.

Chlorinated solvents also may undergo chemical transformation through hydrolysis, losing chlorine atoms and creating a less chlorinated daughter product. This hydrolysis reaction has been observed at sites contaminated with 1,1,1-TCA (D. Bass, Fluor Daniel GTI, unpublished data, 1996).

Remediation Technology Options

Cleaning up chlorinated solvents is significantly more difficult than cleaning up petroleum hydrocarbons. Because the neat solvent, unlike petroleum hydrocarbons, is more dense than water, it can migrate below the water table. Once below the water table, it can penetrate deep into the saturated zone without appreciable spreading and contaminate areas below the water table that are difficult to locate and reach. In addition, chlorinated solvents have a high relative solubility, and their aqueous-phase transport is not significantly slowed by adsorption to aquifer solids. Therefore, a substantial amount of contamination will dissolve from chlorinated solvent source areas and form large plumes of contamination in ground water (Pankow and Cherry, 1996). Despite these difficulties, considerable progress has been made in the past decade in developing and refining methods for cleanup of sites contaminated with chlorinated solvents.

Separation Techniques

Because chlorinated solvents are relatively volatile and soluble, the primary treatment technologies currently used for sites contaminated with chlorinated solvents are separation and extraction processes. The four most widely used technologies for chlorinated solvent cleanup are pump-and-treat systems, SVE, air sparging, and dual-phase extraction (Fluor Daniel GTI, unpublished market survey data, 1996). Because of the high volatility of many solvents, the most efficient of these technologies for dealing with source areas are aeration processes: SVE, air sparging, and dual-phase vacuum extraction. Pump-and-treat systems are effective primarily as containment systems or as a means of treating low concentrations of dissolved contaminants (National Research Council, 1994).

The persistent problems in the treatment of sites contaminated with chlorinated solvents are related to removing the solvents from low-permeability zones and fractured rock and treating residual material and pools of separate-phase DNAPLs. In low-permeability zones, not only is delivering air or water to volatilize or dissolve the contaminant very difficult, but, as discussed above, chlorinated solvents can also penetrate into clays and fractures, making them difficult to find and limiting their extractability. Two approaches are being developed to address these limitations. The first is chemically enhanced removal using surfactants, foams, or cosolvents (Pope and Wade, 1995; Annable et al., 1996; Jafvert, 1996). These processes are designed to desorb, solubilize, or displace residual-phase solvents or DNAPLs. The contaminants are removed through liquid recovery in either an aqueous or nonaqueous phase. The second is thermally enhanced mobilization through the injection of steam (Udell and Stewart, 1989, 1990). These processes have a potential drawback in that they may cause further, unwanted migration of DNAPL. This unintended migration may occur when the remediation processes cause coalescence and/or lowering of the surface tension of the residual-phase material, leading to the formation of a pool of free-product DNAPL that may penetrate deeper into the subsurface (Pennell et al., 1996).

Reaction Techniques

Dissolved chlorinated solvents are somewhat biologically and chemically reactive, and efforts are under way to develop reactive technologies for cleaning up these contaminants. Chemical oxidation is a developing technology that has promise for the direct oxidation of chlorinated ethylenes, which have a carbon-carbon double bond that is vulnerable to oxidative attack. Either ozone or Fenton's reagent (iron-catalyzed H_2O_2) can be used. In addition, chlorinated solvents may be chemically transformed by zero-valent iron (Gillham and O'Hannesin, 1994; Wilson, 1995; Gillham, 1995). The process employs an iron-filled trench (a passive reactive barrier) through which the contaminated ground water flows; the chlorinated solvents are chemically reduced upon contact with the iron (see Box

BOX 3-5
Metallic Iron Barrier for In Situ Treatment of Chlorinated Solvents: Concept and Commercial Application

Over the past several years, numerous laboratory batch and flow-through column experiments have demonstrated that zero-valent iron causes transformation of dissolved chlorinated solvents, such as TCE and PCE, and the reduction and precipitation of chromium (e.g., Gillham and O'Hannesin, 1994; Matheson and Tratnyek, 1994; Powell et al., 1995; Roberts et al., 1996; Burris et al., 1995). The reaction rates for chlorinated solvents depend primarily on the degree of chlorination (highly chlorinated compounds are transformed more rapidly due to favorable energetics) and the reactive surface area of the iron. Relatively rapid rates have been measured, with half lives on the order of a few minutes to hours, for many compounds (Johnson et al., 1996).

The laboratory evidence that iron causes the reduction of chlorinated solvents has been combined with funnel-and-gate systems, a method for directing ground water flow to a reactive treatment zone (Wilson, 1995). The ground water is funnelled through a permeable treatment zone containing iron filings. As the water passes through, the chlorinated solvents are chemically reduced.

The first full-scale in situ permeable iron barrier was installed in Sunnyvale, California, in the fall of 1994 (see Figure 3-3) (ETI, 1995). The contaminated ground water at the site is relatively shallow, and the aquifer is comprised of interfingered sands and sandy silts. The ground water velocity is approximately 0.3 m (1 ft) per day. Treatability studies consisted of laboratory and field column experiments with site ground water containing the primary volatile organic contaminants of concern at the site: cis-1,2-dichloroethene (1,415 µg/liter), TCE (210 µg/liter), and vinyl chloride (540 µg/liter). Vinyl chloride had the longest half life, 4 hours. The reaction rates determined in the treatability studies, combined with information about the ground water flow rate and contaminant concentrations, were used to design the permeable barrier (Yamane et al., 1995).

The reactive zone, comprised of 100 percent reactive iron, is approximately 1.2 m (4 ft) wide and 12 m (40 ft) long and extends between 2 and 6 m (7 and 20 ft) below ground surface. The reactive zone specifications were designed to ensure transformation of the contaminants to less than the cleanup standard. Because vinyl chloride is transformed most slowly and was present at relatively high concentrations at the site, it was the analyte of greatest concern in the design process.

The reactive zone was flanked by slurry walls to direct ground water flow into the zone. A high-permeability zone, comprised of pea gravel, was installed both upgradient and downgradient of the reactive zone to reduce the effects of local heterogeneities on flow through the reactive zone. To date, no chlorinated organic products have been detected in the four downgradient monitoring wells (J. L. Vogan, EnviroMetal Technology, personal communication, 1996).

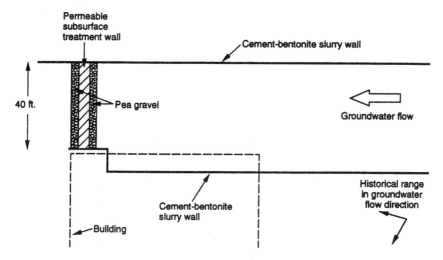

FIGURE 3-3 Plan view of permeable iron barrier for treatment of chlorinated solvents. SOURCE: Reprinted, with permission, from Yamane et al. (1995). © 1995 by American Chemical Society.

3-5). The long-term effectiveness of this approach depends on maintaining a reactive surface free of iron precipitates and biofilms.

Engineered in situ bioremediation of chlorinated solvents has been developing in two areas. The first is the continued study of aerobic cometabolic pathways, in which the bioremediation systems add toluene, natural gas, or propane to the subsurface to stimulate cometabolism of the solvent. This technology was successfully demonstrated at Moffett Naval Air Station in California (Semprini et al., 1990). A second area of development has been the use of sulfate-reducing conditions. This type of bioremediation technology, developed by DuPont (Beeman, 1994), adds sulfate and benzoate to degrade TCE and PCE.

In addition to these engineered forms of bioremediation, considerable research is under way to define anaerobic and aerobic biological processes that lead to intrinsic bioremediation of chlorinated solvents without human intervention (EPA, 1996b). Intrinsic bioremediation of chlorinated solvents in ground water occurs most frequently by reductive dechlorination under anaerobic conditions generated by anthropogenic carbon sources (Chapelle, 1993). Case examples of this are generally scenarios in which chlorinated solvents occur as co-contaminants with anthropogenic sources of dissolved organic carbon, typically hydrocarbon fuels or landfill leachates. Intrinsic bioremediation of chlorinated solvents requires greater than stoichiometric amounts of carbon to serve as a source of energy for the microbes (McCarty, 1996). To date, reports of intrinsic bioremediation of chlorinated solvents are from sites where anthropogenic carbon sources (such as petroleum hydrocarbons) serve as the electron donor (Wiedemeier et al., 1996).

Research Needs

Considerable progress has been made in the last 10 years in treating sites contaminated with chlorinated solvents. However, innovation and development are still needed to improve the efficiency and decrease the costs of existing technologies and to develop new technologies, especially for cleaning up low-permeability zones and chlorinated solvents present as DNAPLs.

Locating source zones of pooled chlorinated solvents poses a major problem, not only for characterizing contaminant distribution at sites but also for designing remediation systems. Costly, extensive sampling often is inadequate to locate chlorinated solvent source zones because of the complexity of DNAPL flow paths and the uncertainties associated with predicting these flow paths.

Research is also needed to improve the scientific basis for designing in situ bioremediation systems for the treatment of chlorinated solvents. A number of laboratory investigations and a few field studies have shown that microbes can degrade chlorinated solvents using various cometabolic pathways and electron acceptors. However, data on process rates and how to control them in situ are insufficient for optimizing remediation system design. Similarly, existing data and models are inadequate for developing accurate predictions of the dynamics of intrinsic remediation.

Chemical or thermal processes that solubilize or mobilize chlorinated solvents require thorough understanding of subsurface fluid movement to ensure that unwanted contaminant migration does not occur. While research has progressed on understanding the physicochemical phenomena that may enhance the mobility of chlorinated solvents, much work is needed to understand how to optimize control of such process fluids in the subsurface.

A final promising area for research related to the treatment of chlorinated solvents is in the assessment of the long-term effectiveness of zero-valent iron barriers for controlling dissolved solvents in ground water. While zero-valent iron barriers are being installed at field sites, the long-term performance of these systems is unknown.

CLEANUP OF POLYCYCLIC AROMATIC HYDROCARBONS

Sources

Polycyclic aromatic hydrocarbon (PAH) compounds are a generally hazardous class of organic compounds found in petroleum and emissions from fossil fuel utilization and conversion processes. PAHs are neutral, nonpolar organic molecules that comprise two or more benzene rings arranged in various configu-

Nonaqueous-phase liquid coal tar aged one year. The coal tar develops an interfacial film that may affect solute dissolution and NAPL wettability characteristics. Courtesy of Richard Luthy, Carnegie Mellon University.

rations. PAHs may also contain alkyl substituents or may be heterocyclic, with the substitution of an aromatic ring carbon with nitrogen, oxygen, or sulfur. Members of this class of compounds have been identified as exhibiting toxic and hazardous properties, and for this reason the EPA has included 16 PAHs on its list of priority pollutants to be monitored in water and wastes.

PAHs are found in process wastes from coal coking, petroleum refining, and coal tar refining and thus may be present in lagoons, sediments, and ground waters that received such products or wastes. Many instances of soil and ground water contamination are reported at facilities where creosote was used for wood treating. Another source of PAH contamination is former manufactured gas plants. Manufactured gas, or town gas, was produced at several thousand such plants. Soil and ground water contamination problems currently exist at many former manufactured gas plants because of prior process operations and residuals management practices (Luthy et al., 1994). Coal tar and associated PAHs are the principal contaminants of concern at these sites.

Fate

The aqueous concentrations of PAHs in natural systems are governed by the hydrophobic character of these compounds and are highly dependent on adsorptive/desorptive equilibria with sorbents present in the system (Dzombak and Luthy, 1984; Means et al., 1980). Also important is whether the PAHs exist as a DNAPL. Because PAHs dissolve only very slowly from DNAPLs, the source of contamination may persist for many years. Indeed, for tar-contaminated soils and

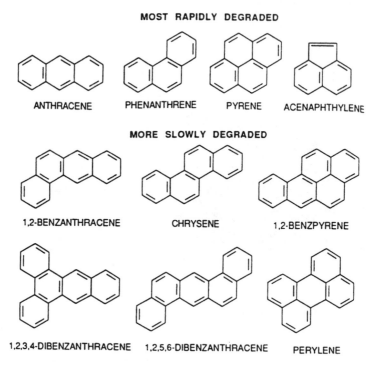

FIGURE 3-4 Relative rates of biodegradation of PAHs in soil. SOURCE: Reprinted, with permission, from Bossert and Bartha (1986).

sediments at manufactured gas plants, the source of PAHs has persisted literally for as long as 100 years (Luthy et al., 1994).

Bacteria, fungi, and algae play important roles in the metabolism of PAHs in terrestrial and aquatic environments (Cerniglia, 1984). Current research indicates that effective microbial degradation of PAHs requires aerobic environments, although microbial degradation of lower-ring PAHs under denitrification conditions has been reported in laboratory studies (Mihelcic and Luthy, 1988). Figure 3-4 shows the relative biodegradability of several PAHs in soil when oxygen is present.

In soils and sediments, the rate of microbial degradation of PAHs may depend on various physicochemical factors affecting the bioavailability of the target compounds to the microorganisms (see Figure 3-5). This is a problem especially with aged and/or weathered samples, which appear to bind PAHs strongly and which often contain a resistant fraction of PAH material that is not amenable to microbial degradation (GRI, 1995; Office of Naval Research et al., 1995; Swiss Federal Institute for Environmental Science and Technology, 1994).

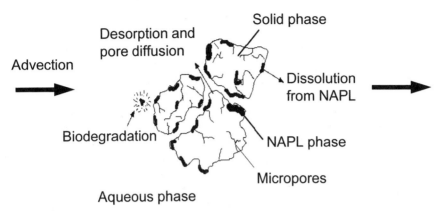

FIGURE 3-5 Illustration of the physicochemical processes in a representative elemental volume that affect the bioavailability of hydrophobic organic compounds for microbial degradation in soil. Contaminants generally degrade when they are dissolved in the bulk ground water. Sorbed contaminants, NAPL-phase contaminants, and contaminants in micropores of solid material are not easily accessible to the microorganisms that cause biodegradation. SOURCE: Reprinted, with permission, from Luthy and Ortiz (1996).

Remediation Technology Options

PAH compounds are relatively persistent in the environment, being resistant to both chemical and microbial transformations. These compounds are not very soluble or volatile, and they tend to sorb to soil surfaces or remain entrapped within an organic phase. Hence, cleanup of PAHs generally focuses on soils and sediments, often ex situ, rather than on contaminants dissolved in ground water. The resistance of PAHs to chemical and microbial transformation, their affinity for soils, and their lack of solubility and volatility make it difficult to treat PAH contamination.

Solidification and Stabilization Techniques

Stabilization/solidification is not commonly the technology of choice for treating soils or sediments contaminated with PAHs or high concentrations of other organic material, although asphalt batching may be appropriate in some instances for tarry matter.

Separation Techniques

Thermal Treatment. Because of the low volatility of PAHs, SVE and other air stripping treatments are not effective remediation techniques for these contaminants. Consequently, separating PAHs from soils requires the use of temperature or chemical solubilizing agents. Thermal destruction using rotary kiln combustion chambers and ex situ thermal treatment to separate volatile and semivolatile contaminants from solids are established technologies and have been used to treat PAHs at full scale (Magee et al., 1994; EPA, 1994c). In a demonstration at a former manufactured gas plant site under the Superfund Innovative Technology Evaluation Program, ex situ thermal desorption was successful in reducing overall PAH levels, but an important conclusion was that materials handling was a significant factor in controlling process performance; the ability to maintain stable desorber operations was linked to soil feed consistency (Maxymillian et al., 1994). In general, most reliability problems occur with material handling, not with the desorption system. Some systems may foul or plug due to the deposition of tar-like material on internal system components. Also, dioxins and furans may be formed during the cleanup process (EPA, 1994c), and this possibility needs to be assessed on a case-by-case basis. An example of an emerging thermal treatment technology is one that employs gas-phase reduction of organic compounds by hydrogen at elevated temperatures; this technology has been tested on harbor sediments containing coal tar (ECO LOGIC, 1995).

Soil Washing and Soil Flushing.   Soil washing without chemical amendments is appropriate for treatment of PAHs in only a few situations, such as in sandy soils having few fines and significant PAH residues associated with separable matter such as wood material (Stinson et al., 1992). Experience with chemical enhancements for soil washing for removal of PAH compounds is limited to a few pilot-scale tests. The few examples include using surfactant with heat (Amiran and Wilde, 1994; EPA, 1994b) and drying and hydrocarbon solvent extraction (Trobridge and Halcombe, 1994). Experience with chemical enhancements for soil flushing is very limited; most information is available from bench-scale tests and a few small field pilot tests. Soil flushing using alkaline reagents, polymer, and surfactant has been pilot tested in a test cell at a wood-treating site (Mann et al., 1993), and an evaluation of in situ steam heating and hot water displacement has been conducted at a former manufactured gas plant site (EPA, 1994c).

Although field experience with chemically enhanced soil flushing and soil washing is limited, considerable research on this topic is under way. One area of research is the use of high concentrations of water-miscible cosolvents, which greatly enhance the solubility of hydrophobic organic contaminants, thereby increasing the mass removal per unit volume of fluid used to flush the contaminated soils (Luthy et al., 1992; Augustijn et al., 1994; Roy et al., 1995). A pilot demonstration of in situ solvent extraction has been conducted at Hill Air Force Base in Utah using ethanol-propanol mixtures injected into gravely sand in a 3 by 5 m test cell having jet fuel as the primary contaminant. In the test, solvent flushing removed on the order of 80 to 90 percent of the hydrocarbon material (Annable et al., 1996). The use of water-miscible solvents has been studied in laboratory tests to evaluate solvent extraction for possible use in cleaning coal tar-contaminated soils. The kinetics of tar dissolution affect treatment duration, although predictions based on laboratory data for a very simple site with no hydrogeologic complexities suggest greater than 90 percent removal in a one-year time frame (Ali et al., 1995).

Another area of research is the use of surfactants that may benefit in situ soil flushing by enhancing the solubility of hydrophobic organic compounds and by lowering the interfacial tension between water and NAPL, resulting in direct mobilization of the NAPL. Surfactant enhancements have been evaluated in various laboratory tests to assess physicochemical phenomena affecting the partitioning of PAH compounds in soil-water systems (Edwards et al., 1991, 1994a,b). Only limited laboratory data are available on how surfactants might affect the PAH transport rate in subsurface environments.

Biological Reaction Techniques

Current understanding of achievable treatment rates and end points for biological treatment of PAHs is very incomplete. Various laboratory tests have described PAH biodegradation in well-controlled systems, but it is unclear how these results translate into understanding of what may occur in field tests. The limited understanding of PAH biodegradation is a particular concern for aged samples from field sites; release of PAHs from aged samples may be much slower than release from freshly applied material. Often, PAHs are completely degradable when freshly applied to soil, but the soil may retain a residual concentration of the same compound after prolonged biological treatment (see Box 3-6). This residual PAH may result from a combination of complex physical and chemical factors controlling solubilization, desorption, and diffusion. Moreover, biodegradation of complex chemical mixtures such as coal tar may be further complicated by substrate interactions causing unpredictable biodegradation patterns (Alvarez and Vogel, 1991), which may include inhibition, competition, and cometabolism.

Various studies have indicated qualitatively that mass transfer limitations may prevent significant biodegradation of PAHs in contaminated soils (Nakles et al., 1991; Morgan et al., 1992; Erickson et al., 1993). Mass transfer limitations could be due to the slow solubilization of PAHs from residual weathered NAPLs or from slow dissolution of PAHs trapped in micropores and sorbed to solid surfaces. As a consequence of these factors, the design of soil treatment systems for PAH compounds requires site-specific laboratory and field tests. Laboratory tests with site samples in aerobic slurry systems may provide an indication of maximum potential biodegradation rates and feasible biotreatment end points.

Bioslurry treatment of PAHs has been evaluated at bench, pilot, and full scale (EPA, 1993b). The bioslurry process uses solids mixing to assist oxygen and nutrient transfer and to enhance mass transfer of solutes and contact with microorganisms. Bench-scale bioslurry treatment of PAHs from creosote-contaminated soil with a 30 percent solids slurry for a 12-week period resulted in

BOX 3-6
Biotreatment of PAHs in Extended Field Trials

The biotreatment of PAHs in field trials often shows a "hockey stick" effect on a plot of PAH concentrations versus time (see Figure 3-6). Total PAH, as represented mainly by 2-, 3-, and 4-ring PAH, may decrease in overall concentration relatively rapidly over several months, but then it often levels off at a residual plateau concentration, which may exceed regulatory limits. Current understanding of PAH biodegradation is insufficient to allow prediction of how this plateau concentration may change with time.

In one study, PAH-contaminated soil was treated in a land treatment field test plot in four lifts from May 1986 through December 1987 at a former creosote wood preserving site (J. Smith et al., 1994). During the period of active biotreatment, involving tillage and nutrient addition, in 1986-1987, total PAH decreased from an initial range of 1,200-3,500 mg/kg to about 800-1,200 mg/kg, with an average of about 50 percent reduction from 2,000 mg/kg. There was noticeable reduction of 2-, 3-, and 4-ring PAH concentrations but hardly any reduction of 5-ring and no reduction of 6-ring PAH concentrations. During the six-year period from 1987-1993, the treated soil was left in place unattended. Sampling in 1993 showed that total PAH had decreased from about 1,000 mg/kg to about 200 mg/kg during the six-year unattended period. Soluble PAH from standard leaching tests was less than 20 µg/liter, with no 4-, 5-, or 6-ring PAH detected and only some 2- and 3-ring PAH (in the parts per billion range) detected. At the end of 1987, the 5- and 6-ring PAH concentrations were about 60-70 mg/kg and 20-30 mg/kg, respectively; in 1993, these values were about half the 1987 concentrations.

These data illustrate that gradual reductions in PAH concentrations may continue at a very slow rate over a number of years in land treatment systems. In addition, as a result of biodegradation and weathering, the remaining contaminants may be much less mobile, as evidenced by field and laboratory leaching assessments.

reduction of total PAHs by 86 percent from initial values of 2,460 mg/kg, with the greatest reduction (more than 98 percent) for 2- and 3-ring PAHs and lower removal rates (72 percent) for the 4-, 5- and 6-ring PAHs (EPA, 1993b).

A pilot-scale demonstration of slurry-phase biotreatment of weathered petroleum sludges was evaluated using a 3.8×10^3 m^3 (1×10^6 gal) reactor retrofitted from a concrete clarifier (EPA, 1993b). The process entailed 56 days of batch operation at about 10 percent solids loading using float-mounted mixers and aerators. Overall, the system reduced PAH concentrations by more than 90 percent.

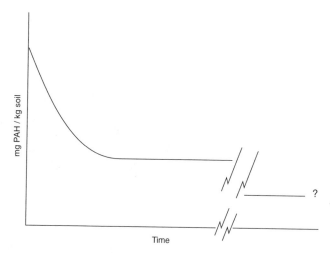

FIGURE 3-6   Typical "hockey stick" pattern observed in degradation of PAHs in soil biotreatment systems.   The biodegradation pattern often exhibits a labile fraction and a resistant fraction; the latter may decrease very slowly over time (see Box 3-6).

Large-scale complete mix pilot reactors are used to evaluate bioslurry treatment technologies.   Courtesy of Remediation Technologies, Inc.

Soil contaminated with coal tar. The residual coal tar may release PAHs at a very slow rate, prolonging biological treatment through lack of adequate bioavailability. Courtesy of Richard Luthy, Carnegie Mellon University.

Greater reductions would probably have been achieved with improved mixing to maintain more solids in suspension. The sludge was a good candidate for bioslurry treatment because it contained mostly 2- and 3-ring PAHs, as opposed to 4-, 5-, and 6-ring PAHs.

An in situ bioslurry system was implemented at full scale to treat wastes at the French Limited site, an abandoned industrial waste lagoon in Harris County, Texas (EPA, 1993e). Process equipment used to optimize oxygenation and contact between microorganisms and the contaminants included mechanical aerators, centrifugal pump sludge mixers, and hydraulic dredge subsoil mixers. Liquid oxygen was injected in pipeline contactors, where it was mixed with the slurry at elevated pressure. This provided more rapid oxygen dissolution with less pumping. The treatment achieved remediation objectives for the compounds, including benzene and benzo(a)pyrene, that were used as indicators of overall contamination.

Research Needs

Additional data are needed to assess thermal treatment processes for PAHs.

Relationships among treatment temperatures, retention times, and overall efficiencies and costs need to be quantified. Materials handling methods need to be optimized.

Various factors remain to be resolved for practical implementation of soil washing and flushing systems that use cosolvents or surfactants for remediation of PAHs. Although laboratory work has been performed to advance the basic science of soil washing and flushing, experience with pilot- or field-scale demonstrations is very limited. Research is needed to improve management of pumped fluids in large-scale applications (including improving delivery of fluids to contaminated zones), evaluate possible reuse of recovered chemicals, and determine the fate of residual contaminants and chemicals remaining in soil.

Although biological treatment of PAHs in soils and sediments has been widely studied, the design of biological PAH treatment systems remains largely empirical. Rate controlling parameters are often unknown. The bioavailability of hydrophobic compounds needs to be enhanced. Degradation rates are highly variable and cannot be predicted reliably. Treatment end points are uncertain. The factors that determine the concentration of PAHs (see Box 3-6 and Figure 3-6) attained after prolonged biotreatment are unknown. The biotreatable fraction of total PAH in a sample is not predictable, nor are the chemical or physical phenomena that may sequester PAHs in soil or sediment known. The effects of aging and weathering on the bioavailability of PAHs cannot be estimated. Data comparing decrease of PAHs by biotreatment with decrease in leachability and toxicity are lacking. The mobility of residual PAHs after biotreatment cannot be predicted. Finally, the ecological effects of residual, relatively insoluble PAHs that may remain after biotreatment are unknown.

CLEANUP OF POLYCHLORINATED BIPHENYLS

Sources

PCB compounds comprise the biphenyl structure with 1 to 10 chlorine atoms, resulting in 209 different structural configurations, or congeners. Each congener has a different number and different positioning of chlorine atoms. PCBs were sold as mixtures of congeners called Aroclors, with each Aroclor having a different weight percent of chlorine. Aroclors were used in a variety of industrial products, including capacitor dielectrics, transformer coolants, heat transfer fluids, plasticizers, and fire retardants in hydraulic oils. Although the use of Aroclors has been banned in many countries, PCBs can be found at low levels dispersed through certain sediment and aquatic systems.

Typically, site contamination problems with PCBs are related to the direct on-site use or disposal of Aroclors at industrial facilities, including the discharge of Aroclors to floor drains, sewers, and lagoons. Many of these disposal practices enhanced migration of the PCBs by providing conduits (drains, sewers, drain fields, boreholes, and wells) to the deeper subsurface. The release of Aroclors to soils also occurred at metal recycling facilities that processed used electrical transformers. There are also instances in which PCB-containing oils were spread on soil and dirt roads for dust and erosion control.

Fate

While each Aroclor mixture is composed of a range of compounds, assessment of the environmental fate and transport of PCBs has been performed typically using average properties of Aroclors (Luthy et al., 1997; Adeel et al., in press). Congener properties, especially solubility and sorption potential, vary widely and strongly affect the fate and transport of PCB compounds (Dzombak et al., 1994). Sorption can significantly retard the movement of PCB compounds and often controls how far the compounds will migrate (Oliver, 1985; Coates and Elzerman, 1986). As with PAHs, the central concern in cleaning up PCB-contaminated sites is the soil, because chemical properties of PCBs limit their migration in ground water.

Remediation Technology Options

The stable chemical properties that made PCBs attractive for use in industrial applications strongly affect remediation technology options. PCBs have very limited solubility in water and are practically nonvolatile. Therefore, containment or stabilization processes are effective for managing PCB-contaminated soils, but separation processes (thermal treatments or chemical extractions) require significant energy inputs. The chemical transformation of PCBs requires elevated temperatures (i.e., incineration or substitution-type processes). Microbial transformations occur very slowly but show promise for toxicity reduction and biostabilization.

Containment and Stabilization Techniques

Various proprietary formulations, consisting of cementing agents or pozzolanic materials, are available in the marketplace for solidification and stabilization of PCB-contaminated soil and sediment. Solidification/stabilization has been used at full scale in an ex situ process to treat PCB-contaminated soil using approximately 30 percent proprietary pozzolanic material and in an in situ process to treat PCB residues using approximately 15 percent calcium oxide and 5 percent kiln dust (Weitzman and Howel, 1989). Technologies for containment of source-area PCBs in the subsurface include slurry trench cut-off walls (made of soil-bentonite, cement-bentonite, or plastic concrete), grout curtains (comprised of cement or chemical grouts), and steel sheet curtains. These barriers are used in conjunction with ground water extraction wells to provide hydraulic containment.

In situ soil mixing is a relatively new technology that avoids excavation by employing a special auger and mixing shaft that permits injection of a slurry consisting of bentonite and water or bentonite and cement. Columbo et al. (1994) and Stinson (1990) summarize an EPA test of a proprietary additive and deep-soil mixing process for an in situ demonstration for stabilizing PCB-contaminated soil. The process decreased PCB mobility by causing ground water to flow around, not through, the monolith. The presence of organic wastes may inhibit the setting and hardening of cement-based or pozzolan-based stabilization technologies.

The EPA has listed one demonstrated in situ vitrification process capable of treating PCBs in soil or sediment (Dávila et al., 1993).

Separation Techniques

Thermal Desorption. PCB-contaminated soils and sediments have been treated in thermal desorbers in field trials (EPA, 1993c). Thermal desorption with reduction of gas-phase PCBs at high temperature with hydrogen has been demonstrated (as part of the EPA Superfund Innovative Technology Evaluation Program) for coal-tar and PCB-spiked harbor sediments (ECI ECO LOGIC, 1992). Material handling, sizing, screening, and conveying often present considerable challenges, depending on the material and type of equipment employed (Lightly et al., 1993). Operation of thermal desorption systems may create up to eight process residual streams requiring attention (Lightly et al., 1993).

Soil Washing and Soil Flushing. As a result of weathering and aging, PCBs may become tightly bound to soils. The physical washing or scrubbing of soil helps to disaggregate the soil matrix and expose the contaminants to the washing media. Soil washing has been demonstrated as effective for separating fine-grained and coarse-grained media (Dávila et al., 1993; EPA, 1993a). In such processes, the fines and humic fractions may be enriched in PCBs, resulting in a smaller volume of material needing subsequent treatment.

There are almost no examples in the literature of field- or pilot-scale tests of soil flushing for PCB removal, except for two tests at an automotive plant site containing PCBs and oils in fill material (Abdul and Ang, 1994; Abdul et al., 1992). Laboratory studies showed that a nonionic alcohol-ethoxylate surfactant could recover more than 80 percent of oil and PCBs from sandy soils. Encouraged by these results, a pilot study was conducted in a 3-m-diameter, 2-m-deep test plot at the automotive plant. The field test employed 0.75 percent aqueous surfactant solution applied with a sprinkler system; the system removed 10 percent (1.6 kg) of the PCB contaminants in 5.5 pore volume displacements. A second test, conducted a year later, employed 2.3 pore volume displacements with 0.75 percent surfactant; in this test, an additional 15 percent (2.5 kg) of the PCB contaminants was removed. The more efficient removal of PCBs in the second test may have been due to an intermittent washing effect, in which the time interval between the first two tests may have allowed for continued diffusion of the contaminant from dead-end pores to the bulk solution. However, subsequent laboratory column test results showed that complete removal of PCBs by surfactant washing would not be practical.

Solvent Extraction.   Solvent extraction has been proven effective for treating soils or sediments containing PCBs (EPA, 1994a). Solvent extraction processes may be divided into three classes, depending on the solvent used: (1) standard, which employs liquid solvents (alkanes, alcohols, and ketones) at near ambient conditions; (2) liquefied gas, which uses gases pressurized at near ambient temperatures; and (3) critical solution temperature, which uses solvents such as triethylamine, which is miscible in water at temperatures less than 18°C (64°F) and only slightly miscible above this temperature. Each of these process types has been evaluated in the field with PCB-contaminated soil or sediment (EPA, 1993d, 1994a, 1995a; Dávila, 1993). Low absolute concentrations of PCBs may not be attained or may require a number of extraction steps.

Reaction Techniques

Chemical.   Substitution processes have been used to treat soils contaminated with PCBs and chlorodibenzodioxins. Although some substitution processes have been available for more than a decade, they have not been used extensively because incineration is cheaper and more widely available and because design problems identified in field tests were not addressed in follow-up work due to lack of funding (Weitzman et al., 1994). Ferguson and Rogers (1990a,b) and GRC Environmental, Inc. (1992) provide technology descriptions and results of some field trials of lower-temperature processes for potassium hydroxide and polyethylene glycol treatments; Friedman and Halpern (1992) provide a description of a process using methoxyethanol and potassium hydroxide. Results of high-temperature substitution processes in field trials with PCB-contaminated soils or sediments are described by Vorum (1991) and EPA (1992) for a reactor employing fuel oil and alkaline polyethylene glycol and Dávila et al. (1993) for treatment using carbonate, hydrocarbon oil, and a catalyst.

Biological.   Laboratory and field monitoring studies indicate that PCBs biodegrade in the environment but at a very slow rate (see Box 3-7). However, work is needed to demonstrate that PCB biodegradation is viable for use in site cleanups (Dávila et al., 1993).

PCB biodegradation occurs through a combination of anaerobic and aerobic microbial processes. Biodegradation under anaerobic conditions can result in reductive dechlorination of highly chlorinated PCBs. As the anaerobic processes progress, the accumulated degradation products may be destroyed aerobically.

Thus, for example, monochlorobenzene that is generated anaerobically from hexachlorobenzene, or likewise the compounds that accumulate in the metabolism of PCBs under anoxic conditions, can be transformed aerobically (Bédard et al., 1987; Mohn and Tiedje, 1992). Such two-stage processes involving an initial anaerobic phase followed by a final aerobic phase represent a promising means for treating PCBs (Alexander, 1994). Harkness et al. (1993) showed in field trials in the Hudson River that lightly chlorinated PCBs were degraded aerobically by native microorganisms when stimulated with oxygen, a mixture of nutrients, and biphenyl to promote cometabolism.

Research Needs

The effectiveness of in situ soil mixing in reducing permeability and stabilizing PCB contaminants is not documented sufficiently. Due to the hydrophobic nature of PCBs, leachability test results are often inconclusive, typically not showing significant differences between treated and untreated material. Furthermore, the effectiveness of soil mixing in thoroughly blending the soil, with no unreacted soil pockets, is not well documented.

To various degrees, substitution reactions convert some of the target molecules to unregulated forms. Although the resultant compounds may be unregulated, the environmental impact of these compounds still needs to be considered. Further proof of the degree of substitution is needed.

Problems with soil washing or flushing include the generation and treatment of large volumes of water; uneven treatment in soil flushing due to nonhomogeneous conditions, including the presence of NAPLs; and the need for improved control of pumped fluids. Problems with surfactant-aided technologies include assessing surfactant losses by degradation or sorption and evaluating surfactant recovery and reuse. Factors affecting the kinetics of surfactant solubilization of PCBs in heterogeneous systems need to be understood to improve process efficiency and chemical use.

Considerable research is needed to understand phenomena affecting the bioremediation of PCB-contaminated soil and sediment in order to properly evaluate the performance of the technology before it can be used for site remediation. In general, the current state of this technology does not permit treatment with confidence at commercial scale. Factors that control the rates of microbial reactions with PCBs, including the coupling of anaerobic and aerobic processes, need to be better understood. The fate and rate of further biodegradation of residual PCBs following active aerobic biological treatment is unknown.

BOX 3-7
Biostabilization of PCBs

Biostabilization refers to the ex situ biodegradation of organic contaminants in soil such that any residual material is not readily released from the soil matrix, or such that residuals are released so slowly that they pose little or no risk to ecological or human health. Following a period of active biological treatment, biostabilized material may be placed in an engineered containment facility and monitored for release of contaminants and to assess whether intrinsic biodegradation processes are adequate to control contaminants released slowly over time.

The Aluminum Company of America is evaluating biostabilization as a method for treatment of sludges and sediments contaminated with PCBs, PAHs, and hydraulic oils (Alcoa Remediation Projects Organization, 1995). Biostabilization is being tested in complete-mix batch slurry bioreactors and in land treatment test plots. The leaching of PCBs before and after treatment is evaluated in laboratory batch leaching tests and in flow-through column tests. The goal of this work is to assess whether aerobic biological treatment with indigenous organisms can substantially reduce the concentrations of potentially mobile, less-chlorinated PCB homologs, thereby permitting placement of treated material in a controlled disposal facility.

Results from an eight-week field test showed 31 to 43 percent overall reduction in PCBs from initial values of 15 to 17 mg/kg of total PCBs based on congener-specific analyses. There was nearly complete re-

CLEANUP OF INORGANIC CONTAMINANTS

Sources

Inorganic contaminants at hazardous waste sites are typically classed as metals[1] (transition or heavy) or radioactive compounds. Some of the most common sources of metal contamination are mine tailing impoundments, plating and smelting operations, and battery recycling plants (National Research Council, 1994). Radioactive contaminants in soil and ground water are a concern primarily at DOE sites as a result of nuclear weapons production. Radioactive elements are also found in nature, but naturally occurring concentrations pose ecological or human health risks in few cases. Although metallic and radioactive contaminants can occur at modest scales, many of the sources of these contaminants, such as

---

[1]The term "metals" is used in this text to refer to transition metals, heavy metals, and radioactive metals and metalloids.

moval of dichloro-PCBs, approximately 60 to 75 percent reduction in the concentration of trichloro-PCBs, and 10 to 15 percent reduction in the concentration of tetrachloro-PCBs, with no significant removal of PCB homologs with 5, 6, or 7 chlorine atoms. The data from laboratory column leach tests showed that aqueous leachate from untreated samples consisted mostly of di-, tri- and tetrachlorobiphenyls. The data from leaching of treated material showed that the concentrations of dichloro-PCBs were substantially reduced, indeed almost eliminated in the land-treated samples, while trichloro-PCBs were removed by about 60 percent, with little change in tetrachloro-PCBs (Adeel et al., in press).

Continued monitoring of the field test plots will occur through subsequent years. Sampling one year later has shown continued decreases in trichloro- and tetrochloro-PCB concentrations. The concentrations of these two homolog groups decreased by about 90 and 55 percent, respectively, from initial levels in the first 460 days of biotreatment (active and passive). Decreased concentrations of more highly chlorinated PCBs were observed during passive biotreatment (J. Smith et al., in press).

Biostabilization is an emerging technology that needs further investigation and development at the laboratory, pilot, and field scales to assess what it may achieve in practice. The concept is being assessed as part of understanding environmentally acceptable end points for soil treatment. More information is needed about the factors controlling the biotransformation, bioavailability, weathering, and release of hydrophobic organic contaminants from soil and sediment in order to provide a stronger underpinning for this technology.

mine tailing impoundments, result in very large sources of potential contamination.

The most commonly detected inorganic and radioactive contaminants in ground water, nitrate and tritium, respectively, are generally not treated (Woodruff et al., 1993). Nitrate occurs in ground water as a result of widespread point and nonpoint sources of pollution, such as farms using nitrogen fertilizers, manure from animal feed lots and pastures, and septic systems. Tritium, a radioactive form of hydrogen, occurs as part of a water molecule and is a concern only at DOE sites. Both nitrate and tritium migrate essentially unretarded in ground water. Nitrate can be treated by osmosis or can serve as an electron acceptor in microbial processes. However, because its occurrence is so widespread and health effects are thought to be limited, ground water restoration is usually not considered, although well-head treatment of drinking water sources is necessary to protect infants from blue baby syndrome. Because tritium is a radioactive element, the only effective way to treat it is to isolate the tritiated water until radioactive decay reduces the concentration to an acceptable level. Because these compounds are not typically of greatest concern at hazardous waste sites (other than DOE sites), they are not discussed in the following text.

Fate

Unlike many organic contaminants, most inorganic contaminants, particularly radioactive ones, cannot be eliminated from the environment by a chemical or biological transformation. Also unlike most organic contaminants, the form of inorganic contaminants significantly affects mobility and toxicity. The form, or speciation, of inorganic contaminants is often determined by the basic geochemistry (e.g., acidity, reduction potential) of the ground water system. Chromium (Cr), for example, is usually present as either Cr(III) (the reduced form), or hexavalent chromium, Cr(VI) (the oxidized form). Cr(VI), which occurs in the mobile anionic forms CrO_4^{2-} and CrO_7^{2-}, is often present in ground water at contaminated sites and is toxic and mobile. In contrast, Cr(III) typically forms relatively insoluble precipitates, which are not readily oxidized and which cause chromium to be relatively permanently immobilized in the environment (Palmer and Wittbrodt, 1991). The inability to eliminate inorganic contaminants by biological or chemical reactions and the strong effect of geochemistry on inorganic contaminant mobility present major challenges in the cleanup of sites containing these contaminants.

For two primary reasons, relatively few metals are soluble and mobile enough to form significant plumes of contamination in typical ground water environments. First, many toxic metals, like chromium, form relatively insoluble carbonate, hydroxide, or sulfide minerals. Precipitation effectively immobilizes the contaminant because the concentration of dissolved contaminant in equilibrium with the precipitate is so low. Formation and dissolution of solid precipitates are controlled primarily by pH, redox conditions, and concentrations of other ions in the ground water. Second, at the near-neutral pH conditions typical of ground water, common hydroxide and silicate mineral surfaces present in aquifers carry a negative charge and thus will strongly sorb many cationic heavy metals by cation exchange, resulting in very low mobility. (If the system is acidic, such as in acidic mine drainage or battery recycling wastes, mineral surfaces typically become positively charged, and cationic metal ions tend not to sorb and to be very mobile.) Thus, most of the metals of greatest concern due to mobility in ground water are present either as anionic (negatively charged) oxides or are present in acidic ground water. An additional concern is the possibility that metals may be transported either by forming complexes with organic matter in the ground water or by sorbing to mobile colloidal particles. Such facilitated transport of metals in ground water is an emerging area of research.

When more than one inorganic contaminant is present at a site, it is important to consider the effect of varying geochemical conditions on the mobility of all the contaminants. Conditions that lower the mobility of one compound may enhance

TABLE 3-5 Speciation and Mobility of Several Inorganic Contaminants (Metals and Radionuclides) of Concern at Hazardous Waste Sites

Dissolved Species	Representative Inorganic Contaminants	Geochemical Conditions Affecting Mobility
Anion or oxyanion	Arsenic (AsO_3^{3-}, AsO_4^{3-}) Chromium (CrO_4^{2-}, $Cr_2O_7^{2-}$) Cyanide (CN^-) Selenium (SeO_3^{2-}, SeO_4^{2-}) Technetium, ^{99}Tc Uranium ($^{234, 235, 238}U$, $UO_2(CO_3)_2^{2-}$, $UO_2(CO_3)_3^{4-}$)	Mobile in moderate to very oxic environments. All oxyanions form relatively insoluble mineral precipitates or coprecipitates, usually with iron and/or sulfide, under very reducing conditions, rendering them relatively immobile in these circumstances. Arsenic can occur in several valences; the most mobile form is AsO_3^{3-}, which occurs under slightly reducing conditions. Uranium can also occur as sulfate complexes and oxide ions.
Cation	Barium (Ba^{2+}) Cadmium (Cd^{2+}) Copper (Cu^+, Cu^{2+}) Lead (Pb^{2+}) Mercury (Hg^+, Hg^{2+}) Nickel (Ni^{2+}) Strontium ($^{90}Sr^{2+}$) Zinc (Zn^{2+})	Cations are mobile in acidic environments. Most of those listed are relatively immobile at moderate to high pH because of the formation of insoluble hydroxide, carbonate, or sulfide minerals. Mercury can form very mobile, highly toxic organic (methyl mercury) complexes in some environments. Strontium mobility is also strongly affected by the presence of calcium, magnesium, and other divalent (2+) cations.

NOTE: Radionuclide isotopes are designated by the isotope number (e.g., ^{90}Sr). Strontium-90 is regulated because it is radioactive, while nonradioactive isotopes of Sr are not regulated.

SOURCES: Hem, 1985; Fetter, 1993; Brookins et al., 1993; L. Smith et al., 1995.

the mobility of another. Table 3-5 indicates the effect of geochemical conditions on the mobility of some of the inorganic contaminants of greatest concern. As shown in the table, species present as cations, such as lead and strontium, are generally mobile only under acidic conditions. Species that are present as oxyanions (oxygen-containing negatively charged species), such as chromium and technetium, are typically relatively mobile in oxic water but form stable precipitates under reducing conditions. Some inorganic contaminants, such as arsenic and mercury, form complexes with organic compounds. Organic complexes tend to be more toxic than the inorganic forms.

Remediation Technology Options: Ground Water

The current standard practices for controlling metal contamination in ground water are to either use a pump-and-treat system to contain the plume or to use institutional controls to restrict human exposure to the contamination. Pump-and-treat systems are not usually effective for plume remediation unless the sources of contamination have been entirely removed. Once extracted, ground water is usually treated by standard water treatment protocols, such as use of pH neutralization, precipitation, flocculation, and sedimentation or reverse osmosis to concentrate and separate the metals into sludge (L. Smith et al., 1995). The sludge must then be disposed of in an appropriate manner. Sludge containing radioactive contaminants can be difficult to dispose of because of cost and lack of adequate disposal facilities.

Because inorganic contaminants cannot be destroyed, innovative technologies focus on either stabilizing the contaminants by decreasing contaminant mobility and toxicity or separating the contaminants from the soil or ground water.

Solidification, Stabilization, and Containment Techniques

In Situ Precipitation and Coprecipitation.   Strategies that exploit precipitation or coprecipitation under reducing conditions are being used or tested for acidic mine drainage water and for mobile oxyanions (L. Smith et al., 1994, 1995). Generally, the goal of these treatments is to immobilize the contaminant in a relatively thermodynamically stable form. In the case of heavy metals in acidic mine drainage, the goal is to precipitate the metals as the reduced sulfide species that were originally present in the mined ore (Wildeman et al., 1994).

At one site, a passive-reactive barrier (permeable treatment wall) for treatment of metal-containing acidic ground water eluting from mine tailings has been operating since 1995. The permeable barrier consists of organic carbon sources (leaf compost, wood chips, and sawdust) mixed with sand to maintain permeability. Within the treatment zone, naturally occurring microorganisms oxidize the carbon source and use the sulfate as the primary electron acceptor. In the process, acid is consumed, neutralizing the pH; reducing conditions are created; and sulfide concentrations are elevated. Metals are sequestered by precipitation as sulfide minerals and by sorption on organic matter. At this site, acidity has decreased (pH has increased from less than 5 to about 7.5), and sulfate concentrations have decreased from about 3,000 mg/liter to less than 10 mg/liter (the detection limit for the experiments) upgradient and downgradient of the wall. Concentrations of iron and nickel, present in the drainage water at 1,000 and 2 mg/liter, respectively, have decreased to less than detectable limits (5 mg/liter and 0.05 mg/liter) in the treatment wall. Theoretically (based on stoichiometric calculations), the mass of carbon in the wall should allow for continued treatment for 20 to 50 years. Although nickel was the only metal at this site present at a concentration greater than the drinking water standard, laboratory column studies have shown that zinc, cadmium, copper, chromium, and cobalt, which also form relatively insoluble sulfide minerals, also may be treatable by this method (D. Blowes, University of Waterloo, personal communication, 1996). Methods for optimizing the carbon sources, reaction rates, and long-term performance of such treatment walls are still under study.

Methods for generating a reducing treatment zone to immobilize oxyanions are also being studied. The approach is to develop reducing conditions in situ such that mobile metals precipitate or coprecipitate as relatively insoluble solid phases; unlike the acidic mine drainage schemes, shifting pH is not a priority for these technologies. Treatment by generating reducing conditions abiotically in a permeable wall containing zero-valent iron is currently being tested at the pilot scale at a site with a plume of chromate-contaminated water from an electroplating facility (Blowes and Ptacek, 1992; Puls et al., 1995; Blowes et al., 1995; Powell et al., 1995). Laboratory studies and initial field results appear promising. The amount of reduced iron installed contains such a large reserve of reducing potential that, theoretically, it should last nearly indefinitely, although research is needed to assess the long-term performance of such systems. The concept is also being considered for treatment of technetium oxide. Long-term field performance is yet to be tested, and optimal techniques for replacing a zone of aquifer material with iron are still evolving.

Other methods for generating a reduced zone in the aquifer by biological or chemical treatment have been laboratory tested but not field tested. Possible strategies include creating an in situ reduced zone in the aquifer by chemically or biologically reducing iron in the sediments, which would result in reduction of contaminants within the treatment zone (DOE, 1994a,b). Such strategies may have an advantage over permeable treatment walls in that they would not require digging up or replacing aquifer solids. However, the methods would have to overcome the potentially large effects of both physical and chemical aquifer heterogeneity and generally would not produce the extreme reducing conditions created by metallic iron.

Geochemically reduced conditions are not favorable for solving all metal contamination problems. Arsenic, for example, can form a more mobile anion in moderately reduced geochemical conditions compared to the species normally present in aerobic ground water. Metals that do not form insoluble solids under reducing conditions would require different treatment methods.

Enhanced Sorption In Situ.   Several methods for enhancing sorption of metals are being tested in the laboratory (DOE, 1994a). The goal of these methods, like in situ precipitation methods, is to immobilize the contaminants. Emplacement of zeolites, immobilized organic chelates, metal-sorbing microorganisms, or iron oxyhydroxide surface coatings on aquifer solids can increase sorption of metals.

Ex Situ Precipitation, Coprecipitation, and Enhanced Sorption.   Over the past decade, constructed wetlands have been increasingly used for treating acidic mine drainage ex situ (Thomson and Turney, 1995) using both aerobic and anaerobic processes (Gusak, 1995). One example is the wetland created at the Big Five

---

BOX 3-8
Created Wetland for Cleanup of Metals

An artificial wetland is successfully treating metals in acidic mine drainage at the Clear Creek/Central City Superfund site near Idaho Springs, Colorado (Morea et al., 1989; Wildeman et al., 1990; Machemer and Wildeman, 1992; Wildeman, 1992; Whiting et al., 1994). The mine drainage at this site has pH less than 3 and high concentrations of zinc (50-70 mg/liter), cadmium (greater than 0.1 mg/liter), and manganese (2-3 mg/liter), as well as some copper and iron. Within the wetland, metals are removed from solution largely by microbiologically mediated precipitation of metal sulfides and by ion exchange on organic matter. Naturally occurring microorganisms use sulfate in the acidic drainage as an electron acceptor, creating an excess of sulfide and neutralizing acidity.

The redox reactions are driven by the carbon source (composted manure) provided to the wetland. In a pilot test at the site, pH increased from 3.0 to greater than 6.5; dissolved concentrations of zinc and copper decreased by more than 98 percent; and dissolved lead and iron concentrations decreased by more than 94 and 86 percent, respectively. The wetland was not effective in removing manganese. Iron removal was variable and depended on seasonal activity of the microorganisms.

Several important operating parameters for designing the full-scale wetland were determined through bench- and pilot-scale testing. The parameters studied included physical, chemical, and microbiological performance of several carbon sources; metals removal mechanisms; requirements for microbiological amendment; and hydraulic requirements (Machemer and Wildeman, 1992; Wildeman, 1992). Pilot-scale testing included design modifications, such as adding baffles to increase contact between the contaminated water and organic matter. Long-term stability of the wetland environment and successful removal of metals is continuing to be studied as full-scale remediation cells are put into place (Whiting et al., 1994).

Constructed wetlands are expected to provide treatment at a fraction of the cost of conventional systems over time and to function much more effectively than conventional systems in remote locations where maintenance is difficult (Wildeman et al., 1990).

Tunnel (see Box 3-8) near Idaho Springs, Colorado. The Big Five Tunnel wetland removes heavy metals from the sulfur-containing water primarily as sulfide precipitates in the anaerobic portion of the wetland (Wildeman et al., 1994). Passive treatment by constructed wetlands is projected to be cost effective relative to lime precipitation for some sites (Gusak, 1995), but the acid neutralization capacity of a wetland can be limited. Wetland treatment basins may need to be cleaned

Field demonstration of the use of created wetlands for the removal of metals from acid mine drainage at a site near Idaho Springs, Colorado (see Box 3-8). Courtesy of Roger Olsen, Camp Dresser & McKee.

periodically (Gusak, 1995). Mining the sludge for the metals may be desirable in some cases.

Electrokinetic Separation Techniques

Use of electrical currents to separate metals from contaminated ground water systems is receiving increasing attention. When a low electrical current is applied to an aqueous system, ions migrate to and are concentrated at the electrodes (Acar and Alshawabkeh, 1993). The reactions can be used to stabilize the metals in situ, or the contaminants can be removed in a concentrated form from the water or process solution surrounding the electrode. Proposed applications focus on ex situ water treatment and in situ or ex situ treatment of fine-grained soils that are difficult to flush because of low permeability. When conducted in situ, the process is similar to dewatering of clays by electroosmosis. Site-specific soil properties control the efficacy of this method. In order to enhance mobility of the target ions, electrode solutions must be added in some cases.

Remediation Technology Options: Soil

The conventional methods for treating soil contaminated with metals include excavation and disposal at an appropriate waste facility, capping the site to prevent infiltration, and institutional controls to reduce exposure to the metals.

At many sites contaminated with metals, such as mine and smelting sites, the volume of contaminated waste solids and soils is so high that removal of the contaminated soil is economically prohibitive. At these sites, the standard treatment is to cap the site to restrict ground water recharge and then to monitor the ground water. Recent efforts at sites where acidic drainage occurs emphasize minimization of contact with the atmosphere (by capping or submerging the tailings) to inhibit acid generation by sulfide mineral oxidation.

For soil contaminated with radioactive substances, the cost of excavation and disposal can be very high. DOE recently estimated that the cost of excavating and disposing of buried transuranic waste from Idaho National Engineering Laboratory was $24,000/m^3 of soil (DOE, 1994a). For some highly radioactive contaminated soils, viable disposal locations may not exist.

Solidification and Stabilization

For representative nonradioactive excavation and disposal applications, solidification and stabilization is a relatively low-cost alternative. Consideration of the metal ion chemistry is essential in producing a material resistant to leaching (Soundararajan, 1992). A disadvantage of this method is the increased waste volume. Additionally, mixtures of metals, which may not be immobilized by the same chemical treatments, can be problematic (L. Smith et al., 1995). The lifetime and/or end use of the stabilized materials must be considered because weathering can potentially remobilize the contaminants (Wiles and Barth, 1992).

Ex situ solidification and stabilization technologies are well established. In situ technologies have not been used extensively but are being developed (L. Smith et al., 1995). A demonstration to treat arsenic-contaminated soil in the San Francisco Bay area with in situ solidification and stabilization was carried out in October 1992 under the Superfund Innovative Technologies Evaluation Program. Post-treatment samples were below the toxicity characteristic leaching procedure arsenic limit of 5 mg/liter for soil with an arsenic concentration ranging from 500 to 5,000 mg/kg.

Because of the potential for remobilization by weathering, the importance of limiting exposure during remediation, and the long half-lives of many radioactive contaminants, many of the standard methods used for solidification and stabilization of heavy metals (see Box 3-3) are inappropriate for radioactive contaminants. As an alternative, the DOE has been developing vitrification techniques, both in situ and ex situ, which were initially used in the nuclear industry as a method for long-term retention of radioactive contaminants. Vitrification is applicable to soils, sludge, or other earthen materials that contain radioactive, inorganic, or organic wastes or waste mixtures. Leach tests have indicated that retention of inorganic and radioactive elements in vitrified material is very good. Factors affecting process performance include the presence of water, large void volumes, or combustible organic material; soil composition; and the electrical conductivity of the waste material. Considerable electrical energy, in the range of 800 to 1,000 kW-hours per ton of soil processed, is required (EPA, 1993e). The technology has been widely tested, primarily at DOE facilities, and has reached the commercialization stage. Cost is generally high (EPA, 1993e). However, for sites at which multiple technologies would otherwise be required or for which there are no other feasible alternatives (i.e., sites containing radioactive waste), this technology may be advantageous.

Biological Reaction

At least three companies are now commercializing phytoremediation systems for the treatment of waste sites contaminated with metals (Watanabe, 1997). Phytoremediation is carried out by growing plants that hyperaccumulate metals in the contaminated soil. The soil is prepared in advance of planting with appropriate amendments, such as chelating agents, to make the contaminants available to the plants. The plants are allowed to grow and accumulate the contaminants and are then harvested. Plant species have been identified that can accumulate zinc, cadmium, lead, cobalt, copper, chromium, manganese, and selenium from contaminated soils. Phytoremediation researchers generally define hyperaccumulators as plants that can store more than 1,000 mg/g of cobalt, copper, chromium, lead, or nickel or 10,000 mg/g of manganese or zinc in their dry matter (Watanabe, 1997). In one field application at a New Jersey industrial site, phytoremediation reportedly restored a site contaminated with 1,000 parts per million of lead during one summer (Watanabe, 1997). Disposal of the harvested plants, especially if they contain high levels of heavy metals, can be a problem. Some believe that the harvested plants may eventually have market value if the metals can be extracted from them and reused, but currently no market for such plants exist.

Separation

Soil washing solutions to remove metals typically contain acids and/or chelating agents, which chemically remove the contaminants from the soil. The process water must be treated before disposal. A number of specific technologies have been tested in recent years (L. Smith et al., 1995). In one demonstration, soil washing with acidification and selective chelation resulted in a two-thirds reduction of lead contamination in the fine fraction of sediment from Toronto Harbor (EPA, 1993e). For inorganic contaminants, soil flushing is not as well developed as soil washing. However, soil flushing has been applied at a few Superfund sites for treatment of chromium, lead, nickel, mercury, and ferrous sulfate (L. Smith et al., 1995).

Research Needs

Radioactive isotopes and mixtures of heavy metals having different chemistries present major challenges in treatment of inorganic contaminants. The only commercially available technology for mixed radionuclides (vitrification) is relatively expensive. Because many technologies focus on immobilizing the contaminants, long-term effectiveness (maintaining the immobile form) is of concern. Most commercially available inorganic contaminant remediation schemes rely on linking several ex situ technologies. Extracting contaminants located at depth in the subsurface is problematic for these technologies. Few technologies are available and tested for treating inorganic contaminants in situ.

CLEANUP OF PESTICIDES

Sources

There are four general classes of pesticides (Grayson and Eckroth, 1985): (1) complex synthetic organics, (2) volatile organics (fumigants), (3) naturally occurring organics, and (4) inorganics. Some of these classes represent a very large number of chemicals. Table 3-6 provides examples of specific pesticides in each class. The following discussion applies primarily to the cleanup of organic pesticides; cleanup of inorganic contaminants is discussed earlier in this chapter.

Pesticide contamination of soil and ground water results from the manufacture, transportation, formulation, and application of herbicides and insecticides. Pesticides are applied as solutions (in water or oil), dusts, or fumigants (vapors). The pure compound or a concentrate is diluted near the point of application to application strength. Often, contamination results from dumping wash waters and residuals from pesticide storage tank cleaning.

The degree of contamination varies as a function of the concentration at which the pesticide was released into the environment. There are generally three types of pesticide contamination scenarios: (1) point-source contamination from pure compounds, (2) point-source contamination from concentrated mixtures, and (3) nonpoint-source contamination from application of the pesticide. In the first case, soil and ground water contamination results from shipping, distributing, and handling the pure compound. In the second case, formulation of the pure compound into dusts or sprays or use of the formulated product (for example at seed treating operations) can result in the release of formulated mixtures. Hydrophobic pesticides are typically formulated with at least one oil and at least two surfactants to allow the chemical to form an emulsion in a water solution and then either

TABLE 3-6 Classes and Uses of Chemical Pesticides

Class	Example	Use
Synthetic organic compounds	Carbamates (propham, aldicarb)	Herbicide, insecticide
	Thiocarbamates (EPTC)	Herbicide
	s-Triazines (atrazine, simazine)	Herbicide
	Dinitroanalines (trifluralin)	Herbicide
	Organosulfur compounds (bentazon, endosulfan)	Herbicide, insecticide
	Phenols (dinoseb)	Herbicide, insecticide
	Organochlorine compounds	
	DDT	Insecticide
	Alachlor	Herbicide
	Cyclodienes (chlordane, heptachlor)	Insecticide
	DCPA	Herbicide
	Chlorinated phenoxy-alkanoic acids (2,4-D)	Herbicide
	Organophosphates (diazanon, malathion)	Insecticide
	Petroleum oils	Insecticide
Fumigants	Ethylene dibromide (EDB)	Insecticide
	Methyl bromide	Insecticide
	Dichlorodibromo propane (DCBP)	Insecticide
	Dichlorodiethyl ether	Insecticide
Naturally occurring organic compounds	Pyrethroids	Insecticide
	Nicotine	Insecticide
	Rotenone	Insecticide
Inorganic compounds	Arsenicals (As_2O_3, $PbHAsO_4$)	Insecticide, herbicide
	Boron compounds	Insecticide, herbicide
	Sulfamates	Herbicide

SOURCE: Grayson and Eckroth, 1985.

stick to the plant leaf or drop to the soil. These formulation chemicals can affect the mobility of the active pesticide ingredients in the environment. The third contamination scenario results from application of the pesticide, whether for agricultural use, golf course maintenance, home lawn care, or other purposes.

Higher contaminant concentrations in ground water and soil result from point sources of pesticides, rather than from nonpoint sources. Proximity to a pesticide formulator, dealer, or applicator has been correlated with high frequency of pesticide detections (Holden et al., 1992). Barbash and Resek (1996) determined that high pesticide concentrations (greater than 100 µg/liter) are not uncommon in ground water beneath agrichemical facilities. Maximum soil concentrations at agrichemical facilities are typically greater than 1,000 µg/kg (Barbash and Resek,

1996). High concentrations of pesticides in ground water have been correlated with proximity to a pesticide distribution plant (Holden et al., 1992).

The frequency of detection of individual pesticides across the United States is most closely related to the frequency of use of the particular pesticide, chemical properties (solubility, volatility, and degradability) of the pesticide, and detection limits of the methods used to analyze for the presence of pesticides. Pesticides consistently detected in several multi-state surveys include the frequently used chemicals atrazine, simazine, alachlor, bentazon, chlordane, dibromochloropro-pane, and ethylene dibromide (Barbash and Resek, 1996). The EPA's National Pesticide Survey, which sampled rural and community ground water supply wells in all 50 states, found as the most frequently detected pesticides dimethyl tetrachloroterephthalate (DCPA) acid metabolites (EPA, 1990). DCPA is widely used for control of broad-leaved weeds and grasses on home and golf course lawns and on farms (EPA, 1990).

Fate

The fate of pesticides is a function of how the pesticide was released to the environment and of the pesticide's chemical properties (solubility, volatility, and degradability). When pure pesticide products are released to the environment, they can behave similarly to organic solvents. Depending on the physical properties of the pesticide, NAPL transport can occur if the pesticide is a liquid or is incorporated in an organic solvent. As with other NAPLs, a portion of the bulk liquid will become entrapped in the soil pores. Subsequent transport processes for the entrapped pesticides include solubilization into ground water and volatilization.

Formulated pesticides that have not been diluted to application strength can behave differently from pure products. The surfactants and solvents used in formulated products can entrain the pesticide and transport it much farther and at higher concentrations than would be predicted based on the solubility of the active ingredient. For example, some pesticides were historically formulated with toluene, a mobile solvent with the potential to transport the pesticide great distances.

Some organic pesticides (such as glyphosate and glufosinate) are relatively soluble and are supplied in a concentrated aqueous solution. When accidentally spilled, these pesticides are transported in aqueous form rather than as an NAPL (although migration of some such pesticides, including glyphosate, may be slowed due to strong sorption by soils). Similarly, pesticides that have been released to the environment as rinse or wash waters enter the environment in dissolved form. As the water solution passes through the soil, some of the dissolved pesticide may sorb to and contaminate the soil. This contaminated soil can then become a long-term source of ground water contamination as the pesticide slowly redissolves.

When formulated pesticides are purposefully applied to a site to control weeds or insects, widespread, low-level contamination of ground water can occur from direct transport to ground water and from leaching of the adsorbed pesticides from the soil into the ground water. Typically, soil fumigants and inorganic pesticides are more soluble and readily leached than more complex and higher-molecular-weight organic pesticides. However, even pesticides that are not readily leached, such as organochlorine and organophosphorus pesticides, can contaminate ground water and surface water (van der Leeden et al., 1990). Because highly insoluble pesticides sorb to soils, they can be transported on soil particles in surface runoff or through the migration of colloidal clay particles in ground water systems.

Volatilization processes can account for significant losses of pesticides applied to the field. Volatilization processes include direct volatilization, wick evaporation, and azeotropic distillation. Under field conditions, the rates of these processes vary widely, both spatially and temporally, depending on soil and pesticide properties and soil environmental conditions (such as soil water content and temperature). In some circumstances, for very volatile pesticides such as methyl bromide and other fumigants, volatilization can account for loss of as much as 80 to 90 percent of the total amount of pesticide applied (Treigel and Guo, 1994). Three competing factors control volatilization processes: (1) the pesticide's vapor pressure and aqueous solubility, (2) the multi-phase distribution of the pesticide in the subsurface, and (3) the water content of the soil (Treigel and Guo, 1994). Biodegradation or other transformation processes can also affect the rate of volatilization. The distribution of pesticides among various phases depends on the sorption coefficient and Henry's law constants as well as the soil water content. Volatile losses are significant primarily for pesticides with very high vapor pressures, low aqueous solubilities, and a very low tendency for sorption. Migration of the pesticide into the subsurface significantly decreases losses due to volatilization.

Once applied to the field, some pesticides will attenuate biologically due to either microbial degradation or plant uptake. The biodegradability of pesticides varies considerably. Naturally occurring organic pesticides are generally biodegraded when the proper nutrients are present. Organophosphate and organonitrogen compounds are often biologically active. Chlorinated compounds such as the cyclodienes and chlorinated aromatics such as dichlorodiphenyltrichloroethane (DDT) are extremely resistant to aerobic degradation. However, many organochlorine pesticides can be partially transformed via reductive dechlorination (an anaerobic process), as in the conversion of DDT to dichlorodiphenyldichloroethane (DDD) and dichlorodiphenylchloroethane (DDE). The byproducts of reductive dechlorination of organochlorine compounds can also be hazardous and quite resistant to further degradation. Byproducts also may be more mobile than the parent compound.

Microbial transformation of pesticides is often much slower in the subsurface than in surface soils, even for pesticides made of natural organic products. A case example is glufosinate ammonium (used to control broad-leaved weeds), which in one examination transformed rapidly in surface soil, with 50 percent disappearance in three to seven days (Gallina and Stephenson, 1992). However, in sandy aquifer sediments in both the laboratory and field settings, glufosinate persisted at high concentrations through three months of experiments (Allen-King et al., 1995). Glufosinate persists because microorganisms use it only as a source of nitrogen, not as a source of carbon and energy, and microbes prefer other sources of nitrogen (such as ammonium) over the glufosinate. Thus, when a competing nitrogen source is present or when carbon is in short supply, glufosinate will not degrade. Other pesticides exhibit similar behavior in that they may be transformed rapidly in warm, high-nutrient surface soil, while transformation rates in ground water are much slower.

Chemical degradation is a more limited pathway but can have an impact on the persistence of pesticides. The primary chemical reactions for many pesticides are hydrolysis, protonation (for amine groups), and oxidation. These reactions may be catalyzed by soils, especially clay minerals. Chemical reactivity is a function of soil pH, moisture, temperature, redox potential, and soil mineralogy.

In summary, pesticide transport properties are extremely complex, perhaps among the most complex of all contaminant groups, and highly variable depending on the type of pesticide, how it entered the environment, and the environmental conditions at the contaminated site.

Remediation Technology Options

The options appropriate for treating pesticide-contaminated soil depend on the nature of the pesticide and the way in which the pesticide was released to the environment. The treatability of pesticides depends on the chemical structure and functional groups of the pesticide because these affect solubility, volatility, degradability, and sorption characteristics. In addition, the surfactants and emulsifiers commonly included in formulated pesticides affect the feasibility of using various remediation technologies.

Conventional approaches to pesticide remediation have been primarily excavation followed by incineration or disposal for contaminated soil and treatment of extracted ground water at the well head. A review of pesticide-contaminated sites in Singhvi et al. (1994) indicates that many alternatives exist for treating pesticide-contaminated soil, but few if any alternatives to conventional pumping and treating have been documented in technical literature for treating pesticide-contaminated ground water. One of the only available case studies reporting on treatment of pesticide-contaminated ground water (Carter et al., 1995) involved ground water and soil remediation at a pesticide processing facility. In this case, dinoseb, metalochlor, volatile organic compounds, and nitrosamine compounds were treated by first excavating and removing the soil and then by pumping the ground water and treating it ex situ with a carbon adsorption and advanced oxidation system. Thus, the following discussion addresses remediation of pesticide-contaminated soil.

Solidification and Stabilization Techniques

Pesticide-contaminated soils and residues can be treated using solidification techniques to reduce contact with water, hence reducing potential dissolution into ground water, or can be altered chemically to reduce mobility. Applications can be performed in situ or (more commonly) ex situ by mixing the soil with a cement-based matrix. Toxicity reduction does not usually occur. This approach is often applied in combination with other pretreatment methods.

Generally, solidification and stabilization techniques have limited applicability to naturally occurring organic compounds with high vapor pressures, complex synthetic organics that form low-concentration plumes in ground water, or organic pesticides that are readily soluble in water. These methods are not preferred for organic or alkylated metal pesticides because the organic fractions may degrade, and the organic ligand can increase the volatility of the metal compound (Singhvi et al., 1994). In other cases, the cementitious materials used to bind the pesticide may not be compatible with the chemical form of the pesticide. Redox-sensitive metals that form soluble oxyanions (such as arsenate and chromate) or become soluble at high pH (such as cadmium hydroxide) would not be appropriate for these methods.

Biological Reaction Techniques

Bioremediation can be applied to treat soil, sludge, and sediments contaminated by halogenated or nonhalogenated organic pesticides that are sufficiently degradable. Bioremediation is not effective for inorganic pesticides containing toxic heavy metals. Ex situ applications may use either a slurry-phase or solid-phase approach. Slurry-phase bioremediation includes mixing of excavated soil or sludge with water and appropriate nutrients in reactor vessels. Generally, biological treatment will not be effective for highly chlorinated pesticides or pesticides present at high concentrations. Degradation of dinoseb and other pesticides in soil from 22 mg/kg to nondetectable levels after 22 days of treatment has been reported in a pilot-scale reactor system (EPA, 1993e).

Two companies[2] are developing coupled anaerobic/aerobic bioremediation as a two-step process for ex situ treatment of pesticide-contaminated soils. In the first step, organochlorine pesticides such as DDT are reductively dechlorinated under strictly anaerobic conditions. In the second stage, the metabolites are degraded aerobically. In most cases, a labile carbon source is added to create an

---

[2]The two companies developing this technology are Zeneca and J. R. Simplot.

anaerobic system. In some cases, water is also used as an oxygen barrier, achieved by saturating the soil, to attain and maintain anaerobic conditions. The aerobic cycle is created by physically turning the soil or mechanically mixing it to aerate it. This process has been shown to degrade DDT, lindane, and methoxychlor (F. Peters, Zeneca Corporation, personal communication, 1996).

Bioremediation of pesticides also may be achieved through land farming and, more recently, composting. During land farming, soil is placed in lined or unlined beds, irrigated, aerated, and supplemented with nutrients as appropriate to maintain biological activity in the soil. Rates of decomposition are pesticide specific and may be influenced by climatic or environmental factors. Below-grade bioremediation cells where the soil is aerated and amended have been specifically designed to treat cyclodiene insecticides (chlordane, heptachlor) and other biodegradable compounds (EPA, 1993e). Preliminary results indicated accelerated degradation and treatment times of between three months and two years. Bicki and Felsot (1994) reported a case study at an Illinois site where soil containing alachlor, atrazine, trifluralin, and metalochlor was treated in experimental plots by land farming. Prolonged persistence of herbicide residues and crop phytotoxicity were potential problems with land application of herbicide-contaminated soil even after 528 days. At a second site, rapid initial degradation of trifluralin after 30 days was followed by slower rates of degradation. Bicki and Felsot (1994) suggest that degradation rates may be related to the length of time herbicide residues are present in the soil and their concentration. Land farming in combination with application of white-rot fungi has also indicated promising results for degradation of pentachlorophenol and other wood treatment wastes (EPA, 1993e).

Composting is conducted by mixing highly contaminated soil with organic matter in piles and providing aeration. Aerobic biological activity causes decomposition of the waste at elevated temperatures within the compost pile.

Intrinsic bioremediation can be an important attenuation route for some types of pesticides, especially for the more mobile fumigants, but many pesticides are quite persistent. Recent research has shown that chlorinated organic compounds composed of carbon chains having two or three carbon atoms can serve as an electron acceptor (Wiedemeier et al., 1995). For such reactions to occur, other degradable organics in addition to the chlorinated organic must be present. The degree of intrinsic degradation of the more complex pesticides is unknown.

Chemical Reaction Techniques

Some pesticides are amenable to oxidation with strong oxidants such as ozone, H_2O_2, Fenton's reagent (H_2O_2 and iron), or potassium permanganate. Some of these oxidants can be enhanced with catalysts and/or ultraviolet radiation. Pesticides can be oxidized in situ through the injection of oxidants, but typically application is ex situ in slurry reactors. Performance is best for aromatic and phenolic compounds and for pesticides (such as cyclodienes) containing unsaturated bonds. Ethers, highly chlorinated compounds, and those with saturated bonds are less amenable to oxidation. The addition of chemical oxidizing agents has been demonstrated at three pesticide-contaminated sites (Singhvi et al., 1994).

Chlorinated pesticides also may be detoxified through removal of the halogenated atoms from the molecule using innovative ex situ treatment processes such as KPEG/APEG (which stands for potassium polyethylene glycol/alkaline polyethylene glycol). This process is applicable to soil containing compounds such as DDT, DDE, lindane, toxaphene, heptachlor, dieldrin, ethylene dibromide, and 2,4-D. It uses alkaline reaction conditions and the reaction of the pesticide with potassium hydroxide polyethylene glycol in the presence of a cosolvent at elevated temperature. This technology is not applicable to inorganic metal pesticides. Newer dechlorination processes use solvated electrons formed by the dissolution of calcium metal in liquid anhydrous ammonia (Abel, 1995).

Separation Techniques

Several separation techniques are applicable to pesticide-contaminated soils. Applicable in situ technologies include soil flushing, SVE, air sparging, steam extraction, and radio frequency heating. These in situ technologies are less frequently applied than ex situ techniques such as those discussed above and thermal desorption (discussed below) because they require more data on the location and distribution of the pesticide to implement and therefore can be more costly than ex situ techniques. Soil flushing, steam extraction, and radio frequency heating are considered emerging technologies.

SVE, possibly combined with air sparging, is applicable to pesticides with vapor pressure greater than 0.5 mm Hg in hydrogeologic settings sufficiently permeable to permit the extraction of vapors. Fumigant compounds (ethylene dibromide, dibromodichloropropane, and methyl bromide), thiocarbamates, and oxide n-acylcarbamates are amenable to this treatment. SVE and air sparging are not applicable to most inorganic and halogenated pesticides because these are typically nonvolatile. SVE has been used to recover pesticides and other volatile compounds at the Sand Creek Superfund site in Colorado (Singhvi et al., 1994).

Ex situ thermal desorption has been selected for at least three Superfund sites and is applicable to volatile and semivolatile compounds, which must be condensed or sorbed in a treatment phase following desorption. Ex situ thermal desorption has demonstrated high removal efficiencies for soils containing toxaphene, DDT, DDE, DDD, endosulfan, dieldrin, endrin, atrazine, diazinon, prometryn, and simazine (Singhvi et al., 1994). Selection of appropriate bed temperatures and residence times is critical in achieving performance standards, particularly with mixtures of pesticide compounds. Volatile metal compounds such as alkyl-mercury or arsenic compounds may also be treated with this technology.

Three in situ techniques are being developed that can treat pesticide-contaminated soils using a combination of heating and vacuum extraction to volatilize and remove the pesticides. Because of the low volatility of many pesticides, the soil often needs to be heated to temperatures at or above the boiling point of water. The first technique used to heat soil and cause contaminant volatilization is electrical resistance heating. Electrodes are placed in the ground, and a current is passed through the soil. Resistance to current flow causes the soil to heat. Simultaneous application of vacuum extraction removes contaminants that volatilize. Electrical resistance heating is effective for attaining temperatures in the range of 80 to 110°C (180 to 230°F) (Heath et al., 1992). The second technique uses radio frequency heating. Antennas (electrodes) are placed in the ground to broadcast microwave energy through the soil. Radio frequency heating can attain temperatures in the range of 250 to 400°C (480 to 750°F) (Roy, 1989). The third heating technique uses radiant heating. A heater element is placed in a steel well. The in-well temperature is raised to approximately 820°C (1500°F), and temperatures of approximately 430°C (810°F) are attained in the sand pack around the well. The wells are placed close enough to attain temperatures of approximately 260°C (500°F) in the surrounding soil. Volatilized contaminants are drawn in through the heated well pack and are partially destroyed (Sheldon, 1996).

Research Needs

The most important research need related to technologies for cleaning up pesticide-contaminated sites is the development of in situ techniques for treatment of contaminated ground water. There is currently essentially a complete absence of experience with technologies other than pump-and-treat systems for treating pesticide-contaminated ground water.

Existing options for treatment of pesticide-contaminated soil need to be optimized and new processes developed. Improved design bases are needed for ex situ biological systems (including coupled anaerobic/aerobic reactors and land farming systems) for treatment of pesticides. Considerable research is also needed to understand the potential for in situ bioremediation of various classes of pesticides. While recent research has demonstrated the potential for intrinsic remediation of pesticides, this research needs to be expanded before the potential for intrinsic remediation to achieve regulatory goals can be accurately predicted. In addition, research is needed to provide a basis for designing in situ bioremediation and phytoremediation systems for treatment of pesticides. Similar work is needed to optimize performance of ex situ chemical reaction systems for treating pesticides and to develop in situ chemical reaction systems. The scientific basis for designing in situ systems that use heat and applied vacuums to remove pesticides also needs to be developed.

An additional area of research applicable to all potential technologies for pesticide remediation is the development of improved pesticide fate and transport models for use in the environmental industry. Many such models have been developed for agricultural purposes, but this knowledge needs to be transferred for application to pesticide remediation. As explained above, pesticides have a wide range of properties; some have high volatility, reactivity, or solubility, while others have very limited volatility, reactivity, or solubility. Effort is needed to organize pesticides into groups of like compounds that can be treated with similar types of remediation technologies.

CONCLUSIONS

From a national perspective, there is little field experience with innovative technologies for treating contaminants other than petroleum hydrocarbons and, to a lesser extent, chlorinated solvents in relatively homogeneous geologic settings. While successes are now quite common in extracting or biologically treating volatile organic compounds in permeable soils, extrapolating such experiences to complex geologic settings, contaminant mixtures, nonaqueous-phase contaminants located at great depth, and contaminants having low volatility or solubility remains a highly uncertain process. For these classes of problems, owners of contaminated sites, the consultants hired to advise them on site remediation, and regulators tend to be risk averse. The vast majority of available technologies are designed for remediation of soils at the surface or in the vadose zone; few options are available for treating contaminated ground water in situ. Research and field work are needed to expand the range of technologies available in the remediation marketplace for treating complex contamination scenarios (see the research recommendations below).

For all types of contamination problems—from sites that are relatively amenable to treatment with existing technologies to the most complex sites—a great deal of uncertainty remains in designing remediation technologies and predicting the results they will achieve. Current designs are often empirical, involving many qualitative assumptions about performance. Research and field work are needed to improve the scientific basis for remediation technology design.

Adding to the uncertainty associated with remediation technology design is lack of information. Obtaining reliable information about innovative remediation technologies is difficult because of the lack of comprehensive data bases, thorough project reports, data collected according to consistent protocols, and peer review of reported data. Much information about remediation technology effectiveness is from literature that has not been peer reviewed. Information and use of technologies at the field scale, especially for sites not covered under the Superfund or underground storage tank programs, is lacking. Data on the success of field trials vary in quality and completeness, and most often such data are not peer reviewed. Methods for determining technology effectiveness and costs vary widely. As a result of the difficulty of obtaining data and the variability of existing data, evaluating and comparing alternative remediation technologies is difficult.

In summary, improving the availability of technologies for cleaning up contaminated sites and the ability to compare these technologies based on rational scientific evaluation will require research, field work, and improved data collection and technology transfer. The development of new remediation approaches will require close links between laboratory and field studies and access to field demonstration sites with the freedom to change process operations during testing. Parallel activities involving field investigations in concert with laboratory and theoretical work will help identify key issues and thus focus scientific and engineering activities on the most critical topics related to remediation.

RECOMMENDATIONS:
TECHNOLOGY INFORMATION AND DISSEMINATION

To improve the quality and availability of data on remediation technology performance, the committee recommends the following actions:

• **The EPA, in collaboration with other stakeholders, should increase the scope and compatibility of data bases containing remediation technology performance information and should make these data bases available on the Internet, with a single World Wide Web page including links to all of the data bases.** Improvements in information collection, assessment, and dissemination are needed to speed development and commercialization of remediation technologies. While a single, centralized data base will likely be unwieldy and may not satisfy the diverse interests of various users, a goal for the EPA should be to help develop comprehensive and electronically accessible data bases that can be readily distributed and manipulated by different contributors and users. These data bases could be established from currently available but incomplete and incompatible systems. To increase portability, a consistent framework for data entry and retrieval should be developed and used in all the data bases. The format for data entry should be simple. The data bases could provide a tiered approach for data entry, with data at the lowest level (consisting of an abstract and short description) not being peer reviewed and data at the highest level having been extensively peer reviewed. The data bases should be widely advertised in peer reviewed and trade journals and at technical conferences so that those who provide data will benefit from increased access to potential technology users.

• **Government agencies, remediation consultants, and hazardous waste site owners should work to increase the sharing of information on remediation technology performance and costs.** Keeping technology performance information proprietary, whether by design or because of lack of dissemination of the information, slows progress in applying the technology elsewhere. Incentives should be developed to encourage submission of technology performance and cost data to the coordinated national data bases recommended above.

• **Government agencies, regulatory authorities, and professional organizations should undertake periodic, comprehensive peer review of innova-**

tive remediation technologies. This type of activity will help define the state of the art, build consensus, and provide a standard for design and implementation of new remediation technologies.

RECOMMENDATIONS: TECHNOLOGY RESEARCH

To expand the range and efficiency of technologies available for cleaning up hazardous waste sites and improve the ability to select and design remediation technologies based on rational scientific analysis, the committee recommends research in the following areas:

• **Hydrogeologic and geochemical phenomena governing contaminant behavior in low-permeability, heterogeneous media.** The fate of contaminants in low-permeability, heterogeneous geologic settings is difficult to predict, and design of remediation systems to access contaminants in these settings is difficult. Improved understanding of contaminant behavior in these complex subsurface systems needs to be coupled with engineering evaluations and rational risk assessment tools to help guide site management decisions.

• **Methods for predicting the fate, effects, and risks of DNAPLs in a wide range of hydrogeologic settings and for removing DNAPLs from the subsurface.** Although long discussed, the problems of locating and treating DNAPL contamination have not been resolved.

• **Treatment of contaminants having limited reactivity and/or mobility, including PAHs, PCBs, pesticides, metals, and radionuclides.** These contaminants are difficult to treat because they can either partially or completely resist destruction by biological or chemical reactions and/or mobilization and extraction from the subsurface. Treatment processes for radionuclides must address the added concern of disposal of extracted radionuclides or risks of radionuclides that are left in place in stabilized or solidified soils.

• **Treatment of contaminant mixtures.** Treatment of contaminant mixtures poses a major challenge because of the variable effects of treatment processes on different types of contaminants. For example, a treatment process that immobilizes, transforms, or degrades one type of contaminant may have little effect on another. Research is needed to identify the types of contaminant mixtures commonly found at hazardous waste sites and the appropriate treatment trains for managing these mixtures. Treatment of mixtures including radioactive contaminants poses a special challenge that needs to be addressed.

• **Factors controlling the bioavailability of hydrophobic organic compounds (especially residual contamination from petroleum hydrocarbons, chlorinated solvents, PAHs, PCBs, and pesticides).** Little information is available on the chemical processes that control slow release and diffusion of compounds through organic contaminant liquids and natural soil organic matter. New investigative methods, conceptual hypotheses, and model frameworks are needed to predict when contaminants are bioavailable, either to microorganisms that can degrade the contaminants or to sensitive human and ecological receptors that may be harmed by the contamination. Information is also needed to determine how biodegradation of hydrophobic organic compounds affects the mobility and toxicity of residuals that remain after active biotreatment. Such information is needed to determine whether contaminants that remain in place but are not bioavailable warrant further remediation.

• **Effectiveness and rates of bioremediation processes, especially those capable of treating organic contaminant mixtures, compounds having low solubilities, strongly sorbed compounds, and compounds resistant to degradation.** Although widely studied, the design of bioremediation processes is, in general, empirical. Rate-controlling parameters are largely unknown, and treatment end points are unpredictable, especially for contaminants other than easily degradable petroleum hydrocarbons.

• **Physical, chemical, and biological processes affecting the rate of intrinsic bioremediation.** As for engineered bioremediation processes, current scientific knowledge is inadequate to provide accurate predictions of the rate and extent of intrinsic bioremediation, especially for contaminants other than easily degradable petroleum hydrocarbons. Such work should address techniques for predicting the rate of intrinsic bioremediation in advance and identifying suitable monitoring strategies for sites where this approach is appropriate.

• **Factors affecting the performance of solvent- and surfactant-based processes for contaminant remediation.** The scientific basis for predicting the performance and kinetics of these processes needs to be improved. Factors related to process performance include hydrologic control of pumped fluids, management and reuse of pumped fluids, doses of solvents and surfactants, effects of residual chemical additives, and heterogeneities in the geologic media.

• **Materials handling for remediation technologies involving mixing and/or moving and processing of large quantities of solids.** Handling of large volumes of soil and sludge can pose equipment and materials problems for both in situ stabilization and solidification techniques and ex situ soil treatment systems. For in situ stabilization and solidification techniques, the ability to achieve desired results requires attention to sampling and geostatistical techniques to ensure the thoroughness of treatment.

• **Long-term effectiveness of in situ solidification, stabilization, and containment techniques.** Having a scientific basis for determining the life of these systems is essential for long-term protection of public health and the environment. Current understanding of the longevity of solidification, stabilization, and containment techniques is inadequate.

• **Long-term effectiveness of in situ biotic and abiotic processes that decrease the mobility of metals.** The effectiveness of novel sorbents for capture of metals needs greater evaluation at the field scale. Soil flushing systems need similar study for metals remediation. Capture of metals by wetlands and other phytoremediation systems appears promising, but many details need to be studied to explain the process and guarantee reliability. Handling mixtures of metals presents substantial problems because of the varying effects of geochemistry on different metals.

REFERENCES

Abel, A. 1995. PCB destruction in soils using solvated electrons. Presented at American Institute Chemical Engineers National Meeting, Boston, August 1, 1995.

Abdul, A., and C. C. Ang. 1994. In situ surfactant washing of polychlorinated biphenyls and oils from a contaminated field site: Phase II pilot study. Ground Water 32 (September-October):727-734.

Abdul, A. S., T. L. Gibson, C. C. Ang, J. C. Smith, and R. E. Sobczynski. 1992. In situ surfactant washing of polychlorinated biphenyls and oils from a contaminated site. Ground Water 30(March-April):219-231.

Acar, Y. B., and A. N. Alshawabkeh. 1993. Principles of electrokinetic remediation. Environmental Science & Technology 27:2638-47. ·

Adeel, Z., R. G. Luthy, D. A. Dzombak, S. B. Roy, and J. R. Smith. In press. Leaching of PCB compounds from untreated and biotreated sludge-soil mixtures. Journal of Contaminant Hydrology.

Aelion, C. M., and P. M. Bradley. 1991. Aerobic biodegradation potential of subsurface microorganisms from a jet fuel-contaminated aquifer. Applied and Environmental Microbiology 57:57-63.

Alcoa Remediation Projects Organization. 1995. Bioremediation Tests for Massena Lagoon Sludges/ Sediments, Vol. 1. Report prepared for the U.S. EPA and New York State Department of Environmental Conservation. Pittsburgh, Pa.: Aluminum Company of America.

Alexander, M. 1994. Biodegradation and Bioremediation. San Diego: Academic Press.

Ali, M. A., D. A. Dzombak, and S. B. Roy. 1995. Assessment of in situ solvent extraction for remediation of coal tar sites: Process modeling. Water Environment Research 67:16-24.

Allen-King, R. M., J. F. Barker, R. W. Gillham, and B. K. Jensen. 1994a. Substrate and nutrient limited toluene biotransformation in sandy soil. Environmental Toxicological Chemistry 13:693-705.

Allen-King, R. M., K. E. O'Leary, R. W. Gillham, and J. F. Barker. 1994b. Limitations on the biodegradation rate of dissolved BTEX in a natural unsaturated, sandy soil: Evidence from field and laboratory experiments. Pp. 175-191 in Hydrocarbon Bioremediation, R. E. Hinchee, B. C. Alleman, R. E. M. Hoeppel, and R. N. Miller, eds. Ann Arbor: Lewis Publishers.

Allen-King, R. M., B. J. Butler, and B. Reichert. 1995. Fate of the herbicide glufosinate ammonium in the sandy, low organic-carbon aquifer at CFB Borden, Ontario, Canada. Journal of Contaminant Hydrology 18:161-179.

Alvarez, P., and T. Vogel. 1991. Substrate interactions of benzene, toluene, and para-xylene during microbial degradation by pure cultures and mixed culture aquifer slurries. Applied and Environmental Microbiology 57(10):2891-1985.

Amiran, M. C., and C. L. Wilde. 1994. PAH removal using soil and sediment washing at a contaminated harbor site. Remediation (Summer):319-330.

Annable, M. D., P. S. C. Rao, R. K. Sillan, K. Hatfield, W. D. Graham, A. L. Wood, and C. G. Enfield. 1996. Field-scale application of in-situ cosolvent flushing: Evaluation approach. In Proceedings of ASCE Conference on NAPLs, Washington, D.C., November 10-14, 1996. New York: American Society of Civil Engineers.

API (American Petroleum Institute). 1996. Petroleum Industry Environmental Performance. Washington, D.C.: American Petroleum Institute.

Armstrong, A. Q., R. E. Hodson, H. M. Hwang, and D. L. Lewis. 1991. Environmental factors affecting toluene degradation in ground water at a hazardous waste site. Environmental Toxicological Chemistry 10:147-158.

Augustijin, D. C. M., R. E. Jessup, P. S. C. Rao, and A. L. Wood. 1994. Remediation of contaminated soils by solvent flushing. ASCE Journal of Environmental Engineering 120(1):42-57.

Barbaro, J. R., J. F. Barker, L. A. Lemon, and C. I. Mayfield. 1992. Biotransformation of BTEX under anaerobic, denitrifying conditions: Field and laboratory conditions. Journal of Contaminant Hydrology 11:245-272.

Barbash, J. E., and E. R. Resek. 1996. Pesticides in Ground Water: Distribution, Trends and Governing Factors. Chelsea, Mich.: Ann Arbor Press.

Bass, D., and R. Brown. 1996. Air sparging case study data base update. Presented at First International Symposium on In Situ Air Sparging for Site Remediation, Las Vegas, Nevada, October 26-27, 1996.

Bédard, D. L., R. E. Wagner, M. J. Brennan, M. L. Haberl, and J. F. Brown, Jr. 1987. Applied and Environmental Microbiology 535:1094-1102.

Beeman, R. E. 1994. In Situ Biodegradation of Ground Water Contaminants. U.S. Patent Number 5, 277, 815. Washington, D.C.: U.S. Patent and Trademark Office

Bianchi-Mosquera, G. C., R. M. Allen-King, and D. M. Mackay. 1994. Enhanced degradation of dissolved benzene and toluene using a solid oxygen-releasing compound. Ground Water Monitoring and Remediation 9(1):120-128.

Bicki, T. J., and A. S. Felsot. 1994. Remediation of pesticide contaminated soil at agrichemical facilities. In Mechanisms of Pesticide Movement into Ground Water, R. C. Honeycutt and D. J. Schabacher, eds. Boca Raton, Fla.: CRC Press.

Blowes, D. W., and C. J. Ptacek. 1992. Geochemical remediation of groundwater by permeable reactive walls: Removal of chromate by reaction with iron-bearing solids. Presented at Subsurface Restoration Conference, Third International Conference on Groundwater Quality Research, Dallas, Texas, June 21-24, 1992.

Blowes, D. W., C. J. Ptacek, C. J. Hanton-Fong, and J. L. Jambor. 1995. In-situ remediation of chromium contaminated groundwater using zero-valent iron. Pp. 780-784 in Preprints of Papers Presented at the 209th American Chemical Society National Meeting, Anaheim, Calif., April 2-7, 1995. Washington, D.C.: American Chemical Society, Division of Environmental Chemistry.

Bossert, I. A., and R. Bartha. 1986. Structure-biodegradability relationships of polycyclic aromatic hydrocarbons in soil. Bulletin of Environmental Contamination and Toxicology 37:490-495.

Brookins, D. G., B. M. Thomson, P. A. Longmire, and P. G. Eller. 1993. Geochemical behavior of uranium mill tailing leachate in the subsurface. Radioactive Waste Management and the Nuclear Fuel Cycle 17(3-4):269-287.

Brown, R. A. 1992. Air Sparging: A Primer for Application and Design. Trenton, N.J.: Fluor Daniel GTI.

Brown, R. A., and F. Jasiulewicz. 1992. Air sparging: A new model for remediation. Pollution Engineering (July):52-55.

Brown, R. A., and R. D. Norris. 1986. An in-depth look at bioreclamation. Presented at HazMat, Atlantic City, N.J., June 2-4, 1986.

Brown, R. A., and R. Falotico. 1994. Dual phase vacuum extraction systems: Design and utilization. Presented at Twenty-sixth Mid-Atlantic Industrial and Hazardous Waste Conference, Newark, Del., August 7-9, 1994.

Brown, R. A., E. L. Crockett, and R. D. Norris. 1987a. The principles of in situ biological treatment. Presented at HazMat West, Long Beach, Calif., December 3-5, 1987.

Brown, R. A., G. E. Hoag, and R. D. Norris. 1987b. The remediation game: Pump, dig or treat. Presented at Water Pollution Control Federation Conference, Philadelphia, October 5-8, 1987.

Brown, R. A., J. Crosby, and R. D. Norris. 1990. Oxygen sources for in situ bioreclamation. Presented at Water Pollution Control Federation Conference, Washington, D.C., December 1990.

Brown, R. A., W. Mahaffey, and R. D. Norris. 1993. In situ bioremediation: The state of the practice. Pp. 121-135 in In Situ Bioremediation: When Does It Work? Washington, D.C.: National Academy Press.

Burris, D. R., T. J. Campbell, and V. S. Manoranjan. 1995. Sorption of trichloroethylene and tetrachloroethylene in a batch reactive metallic iron-water system. Environmental Science & Technology 29:2850-2855.

Carter, R. W., H. Stiebel, P. J. Nalasco, and D. L. Pardieck. 1995. Investigation and remediation of groundwater contamination at a pesticide facility: A case study. Water Quality Research Journal Canada 30(3):469-491.

Cerniglia, C. E. 1984. Microbial metabolism of polycyclic aromatic hydrocarbons. Advances in Applied Microbiology 30:31-71.

Chapelle, F. H. 1993. Ground-Water Microbiology and Geochemistry. New York: John Wiley & Sons.

Claus, D., and N. Walker. 1964. The decomposition of toluene by soil bacteria. Journal of General Microbiology 36:107-122.

Coates, J. T., and A. W. Elzerman. 1986. Desorption kinetics for selected PCB congeners from river sediments. Journal of Contaminant Hydrology 1:191-210.

Cohen, R. M., and J. W. Mercer. 1993. DNAPL Site Evaluation. Boca Raton, Fla.: C. K. Smoley.

Columbo, P., E. Barth, P. Bishop, J. Buelt, and J. R. Connor. 1994. Stabilization/Solidification. Vol. 4 of Innovative Site Remediation Technology, W. C. Anderson, ed. Annapolis, Md.: American Academy of Environmental Engineers.

Dávila, B., K. W. Whitford, and E. S. Saylor. 1993. Technology Alternatives for the Remediation of PCB-Contaminated Soil and Sediment. EPA/540/S-93/506. Washington, D.C.: EPA, Office of Solid Waste and Emergency Response.

Delta Omega Technologies. 1994. Creo-Solv Technical Report: Soil Washing Applications for Creosote Removal. DOT-CS-1601. Houston, Tex.: Delta Omega Technologies.

DOD (Department of Defense) Environmental Technology Transfer Committee. 1994. Remediation Technologies Screening Matrix and Reference Guide, Second Edition. NTIS PB95-104782. Springfield, Va.: National Technical Information Service.

DOE (Department of Energy). 1994a. In Situ Remediation Integrated Program: Technology Summary. DOE/E0134P. Springfield, Va.: National Technical Information Service.

DOE. 1994b. Technology Catalogue. Springfield, Va.: National Technical Information Service.

Dzombak, D. A., and R. G. Luthy. 1984. Estimating Sorption of Polycyclic Aromatic Hydrocarbons on Soils. Soil Science 137:292-308.

Dzombak, D. A., R. G. Luthy, Z. Adeel, and S. B. Roy. 1994. Modeling Transport of PCB Congeners in the Subsurface. Report to the Aluminum Company of America, Environmental Technology Center, Alcoa Center, Pa. Pittsburgh, Pa.: Carnegie Mellon University, Department of Civil and Environmental Engineering.

ECI ECO LOGIC International, Inc. 1992. Pilot-Scale Demonstration of Contaminated Harbor Sediment Treatment Process: Final Report. Rockwood, Ontario: ECI ECO LOGIC.

ECO LOGIC. 1995. The ECO LOGIC Process: A Gas-Phase Chemical Reduction Process for PCB Destruction. Rockwood, Ontario: ECO LOGIC Corporation.

Edwards, D. A., R. G. Luthy, and Z. Liu. 1991. Solubilization of polycyclic aromatic hydrocarbons in micellar nonionic surfactant solutions. Environmental Science & Technology 25:127-133.

Edwards, D. A., Z. Adeel, and R. G. Luthy. 1994a. Distribution of nonionic surfactant and phenanthene in sediment/aqueous systems. Environmental Science & Technology 28:1550-1560.

Edwards, D. A., Z. Liu, and R. G. Luthy. 1994b. Surfactant solubilization of organic compounds in soil/aqueous systems. ASCE Journal of Environmental Engineering 120:5-22.

Ely, D. L., and D. A. Heffner. 1991. Process for In Situ Biodegradation of Hydrocarbon Contaminated Soil. U.S. Patent 5,017,289. Washington, D.C.: U.S. Patent and Trademark Office.

EPA. 1990. National Pesticide Survey: Summary Results of EPA's National Survey of Pesticides in Drinking Water Wells. Washington, D.C.: EPA.

EPA. 1992. Demonstration Bulletin: AOSTRA—SoilTech Anaerobic Thermal Processor: Wide Beach Development Site, Superfund Innovative Technology Evaluation. EPA/540/MR-92/008. Cincinnati, Ohio: EPA, Risk Reduction Engineering Laboratory.

EPA. 1993a. Applications Analysis Report: Bergmann USA Soil/Sediment Washing Technology, Preliminary Draft. Cincinnati, Ohio: EPA, Risk Reduction Engineering Laboratory.

EPA. 1993b. Pilot-Scale Demonstration of a Slurry-Phase Biological Reactor for Cresote Contaminated Soil: Applications Analysis Report. EPA/540/A5-91/009 Washington, D.C: EPA, Office of Research and Development.

EPA. 1993c. Demonstration Bulletin: X*TRAX Model 200 Thermal Desorption System, Superfund Innovative Technology Evaluation. EPA/540/MR-93/502. Cincinnati, Ohio: EPA, Risk Reduction Engineering Laboratory.

EPA. 1993d. Resources Conservation Company B.E.S.T. Solvent Extraction Technology, Application Analysis Report. EPA/540/AR-92/079. Washington, D.C.: EPA, Office of Research and Development.

EPA. 1993e. Superfund Innovative Technology Evaluation Program Technology Profiles, Sixth Edition. EPA/540/R-93/526. Washington, D.C.: EPA.

EPA. 1994a. Engineering Bulletin—Solvent Extraction. EPA/540/S-94/503. Washington, D.C.: EPA, Office of Emergency and Remedial Response.

EPA. 1994b. Status Reports on In Situ Remediation Technologies for Ground Water and Soils at Hazardous Waste Sites: Surfactant Enhancements (Draft). EPA 542-K-94-003. Washington, D.C.: EPA, Office of Solid Waste and Emergency Response.

EPA. 1994c. Status Reports on In Situ Remediation Technologies for Ground Water and Soils at Hazardous Waste Sites: Thermal Enhancements (Draft). EPA 542-K-94-009. Washington, D.C.: EPA, Office of Solid Waste and Emergency Response.

EPA. 1995a. In Situ Remediation Technology Status Report: Surfactant Enhancements. EPA 542-K-94-003. Washington, D.C.: EPA.

EPA. 1995b. In Situ Remediation Technology Status Report: Thermal Enhancements. EPA-542-K-94-009. Washington, D.C.: EPA.

EPA. 1996a. Innovative Treatment Technologies: Annual Status Report (Eighth Edition). EPA-542-R-96-010. Washington, D.C.: EPA, Office of Solid Waste and Emergency Response.

EPA. 1996b. Symposium on Natural Attenuation of Chlorinated Organics in Ground Water, Dallas, September 11-13, 1996. EPA540/R-961509. Washington, D.C.: EPA, Office of Research and Development.

Erickson, D. C., R. C. Loehr, and E. F. Neuhauser. 1993. PAH loss during bioremediation of manufactured gas plant soils. Water Research 27:911-919.

ETI (EnviroMetal Technology Inc.). 1995. Performance history of the envirometal process, Internal document. October 1995.

Federal Remediation Technologies Roundtable. 1995a. Accessing Federal Data Bases for Contaminated Site Clean-Up Technologies. Washington, D.C.: EPA.

Federal Remediation Technologies Roundtable. 1995b. Guide to Documenting Cost and Performance for Remediation Projects. EPA-542-B-95-002. Washington, D.C.: EPA.

Ferguson, T. L., and C. J. Rogers. 1990a. Field Applications of the KPEG Process for Treating Chlorinated Wastes: Project Officers Report. PB89-212-724/AS. Springfield, Va.: National Technical Information Service.

Ferguson, T. L., and C. J. Rogers. 1990b. Comprehensive Report on the KPEG Process for Treating Chlorinated Wastes. EPA/600/S2-90/026. Cincinnati, Ohio: Risk Reduction Engineering Laboratory.

Fetter, C. W. 1993. Contaminant Hydrogeology. New York: Macmillan.

Friedman, A. J., and Y. Halpern. 1992. Untreated and biotreated sludge-soil mixtures. Journal of Contaminant Ecology.

Gallina, M. A., and G. R. Stephenson. 1992. Dissipation of [^{14}C]glufosinate ammonium in two Ontario soils. Journal of Agricultural and Food Chemistry 40:165-168.

Gillham, R. W. 1995. Resurgence in research concerning organic transformations enhanced by zero-valent metals and potential application in remediation of contaminated groundwater. Pp. 691-694 in Preprints of Papers presented at the 209th American Chemical Society National Meeting, Anaheim, Calif., April 2-7, 1995. Washington, D.C.: American Chemical Society, Division of Environmental Chemistry.

Gillham, R. W., and S. F. O'Hannesin. 1994. Enhanced degradation of halogenated aliphatics by zero-valent iron. Ground Water 32:958-967.

Gotlieb, I., J. W. Bozzelli, and E. Gotlieb. 1993. Soil and water decontamination by extraction with surfactants. Separation Science and Technology 28(1-3):793-804.

Grayson, M., and D. Eckroth. 1985. Kirk-Othmer Concise Encyclopedia of Chemical Technology. New York: John Wiley & Sons.

GRC Environmental, Inc. 1992. Alkaline Dechlorination Using Dimethyl Sulfoxide. East Syracuse, N.Y.: GRC Environmental, Inc.

GRI (Gas Research Institute). 1995. Proceedings of Workshop on Environmentally Acceptable Endpoints in Soil, Washington, D.C., May 4-5, 1995. Chicago, Ill.: GRI.

Grubb, D. G., and N. Sitar. 1994. Evaluation of Technologies for In Situ Cleanup of DNAPL Contaminated Sites. EPA/600/R-94/120. Ada, Okla.: EPA, R. S. Kerr Environmental Research Laboratory.

Gusak, J. J. 1995. Passive-treatment of acid rock drainage: What is the potential bottom line? Mining Engineering 47(3):250-253.

GWRTAC (Ground-Water Remediation Technologies Analysis Center). 1995. NETAC selected to operate national ground-water remediation technology center (news release). Pittsburgh, Pa.: GWRTAC (http://www.chmr.com/gwrtac).

Harkness, M. R., J. B. McDermott, D. A. Abramowicz, J. J. Salvo, W. P. Flanagan, M. L. Stephens, F. J. Mondello, R. J. May, J. H. Lobos, K. M. Carroll, M. J. Brennan, A. A. Bracco, K. M. Fish, G. L. Warner, P. R. Wilson, D. K. Dietrich, D. T. Lin, C. B. Morgan and W. L. Gately. 1993. In situ stimulation of aerobic PCB biodegradation in Hudson River sediment. Science 259:503-507.

Heath, W. A., J. S. Roberts, D. L. Lenor, and T. M. Bergman. 1992. Engineering scale-up of electrical soil heating for soil decontamination. Presented at Spectrum '92, Boise, Idaho, August 23-27, 1992. Richland, Wash.: Pacific Northwest Laboratory.

Hem, J. D. 1985. Study and Interpretation of the Chemical Characteristics of Natural Water. U.S. Geological Survey Water-Supply Paper 2254, Third Edition. Washington, D.C.: Government Printing Office.

Hinchee, R. E., D. C. Downey, and T. Beard. 1989. Enhancing biodegradation of petroleum hydrocarbons through soil venting. In Proceedings, Petroleum Hydrocarbons and Organic Chemicals in Ground-Water: Prevention, Detection, and Restoration, Houston, November 5-7, Houston. Worthington, Ohio: National Water Well Association.

Hinchee, R.E., and S. K. Ong. 1992. A rapid in situ respiration test for measuring aerobic biodegradation rates of hydrocarbon in soil. Journal of Air and Waste Management 42(10):1305.

Hoag, G. E., and C. Bruel. 1988. Use of soil venting for treatment/reuse of petroleum contaminated soil. Pp. 301-306 in Soil Effects, E. J. Calabrese and P. T. Kostecki, eds. New York: John Wiley & Sons.

Holden, L. R., J. A. Graham, R. W. Whitmore, W. J. Alexander, R. W. Pratt, S. K. Liddle, and L. L. Piper. 1992. Results of the national alachlor well water survey. Environmental Science & Technology 26:935-943.

Hutchins, S. R., and J. T. Wilson. 1994. Nitrate-based bioremediation of petroleum-contaminated aquifer at Park City, Kansas: Site characterization and treatability study. Pp. 80-92 in Hydrocarbon Bioremediation, R. E. Hinchee, B. C. Alleman, R. E. M. Hoeppel, and R. N. Miller, eds. Ann Arbor, Mich.: Lewis Publishers.

Jafvert, C. T. 1996. Surfactants/Cosolvents. Technology Evaluation Report TE-96-62. Pittsburgh, Pa.: Ground Water Remediation Technologies Analysis Center.

Johnson, P. C., and R. A. Ettinger. 1994. Considerations for the design of in situ vapor extraction systems: Radius of influence vs. radius of remediation. Groundwater Monitoring and Remediation 14(3):123-138.

Johnson, P. C., C. Stanley, M. Keblowski, D. Byers, and J. Corhart. 1990. A practical approach to design, operations, and monitoring of in situ soil-venting systems. Ground Water Monitoring Review 10(2):159.

Johnson, R. L., W. Bagby, M. Perrott, and C. Chen. 1992. Experimental examination of integrated soil vapor extraction techniques. In Proceedings of Petroleum Hydrocarbons and Organic Chemicals in Ground Water: Prevention, Detection, and Restoration. Dublin, Ohio: National Ground Water Association.

Johnson, T. J., M. M. Scherer, and P. G. Tratnyek. 1996. Kenetics of halogenated organic compound degradation by iron metal. Environmental Science & Technology 30:2634-2640.

Kittel, J. A., A. Leeson, R. E. Hinchee, R. N. Miller, and P. Haas. 1995. Results of a multisite treatability test for bioslurping: A comparison of LNAPL rates using vacuum-enhanced recovery (bio-slurping), passive skimming, and pump drawdown recovery techniques. In Proceedings, Organic Chemicals in Groundwater: Detection, Prevention, and Remediation. Dublin, Ohio: National Ground Water Association.

Leder, A., and Y. Yoshida. 1995. C_2 Chlorinated Solvents (www-cmrc.sru.com/CIN/mar-apr95/article10.html). Menlo Park, Calif.: SRI International.

Lightly, J., M. Choroszy-Marshall, M. Cosmos, V. Cundy, and P. DePercin. 1993. Thermal Desorption. Vol. 6 of Innovative Site Remediation Technology, W. C. Anderson, ed. Annapolis, Md.: American Academy of Environmental Engineers.

Luthy, R. G., D. A. Dzombak, C. A. Peters, M. A. Ali and S. B. Roy. 1992. Solvent Extraction for Remediation of Manufactured Gas Plant Sites. Final Report TR-10185, Research Project 3072-02. Palo Alto, Calif.: Electric Power Research Institute.

Luthy, R. G., D. A. Dzombak, C. A. Peters, S. B. Roy, A. Ramaswami, D. V. Nakles, and B. R. Nott. 1994. Remediating tar-contaminated soils at manufactured gas plant sites. Environmental Science & Technology 28:266A-276A.

Luthy, R. G., and E. Ortiz. 1996. Bioavailability and biostabilization of hydrophobic organic compounds. Paper presented at UIB-GBR-CSIC-TUB Symposium on Biodegradation of Organic Pollutants, Mallorca, Spain, June 29-July 3, 1996.

Luthy, R. G., D. A. Dzombak, M. Shannon, R. Utterman, and J. R. Smith. 1997. Aqueous solubility of PCB congeners from an Aroclor and an Aroclor/hydraulic oil mixture. Water Research 31(3):561-573.

Lyman, W. J., P. J. Reidy, and B. Levy. 1992. Mobility and Degradation of Organic Contaminants in Subsurface Environments. Chelsea, Mich.: C. K. Smoley.

Machemer, S. D., and T. R. Wildeman. 1992. Adsorption compared with sulfide precipitation as metal removal processes from acid mine drainage in a constructed wetland. Journal of Contaminant Hydrology 9(112):115-131.

Magee, R. S., J. Cudahy, C. R. Dempsey, J. R. Ehrenfeld, F. W. Holm, D. Miller, and M. Modell. 1994. Thermal Destruction. Vol. 7 of Innovative Site Remediation Technology, W. C. Anderson, ed. Annapolis, Md.: American Academy of Environmental Engineers.

Mann, M. J., D. Dahlstrom, P. Esposito, L. Everett, G. Peterson, and R. P. Traver. 1993. Soil Washing/Soil Flushing. Vol. 3 of Innovative Site Remediation Technology, W. C. Anderson, ed. Annapolis, Md.: American Academy of Environmental Engineers.

Matheson, L. J., and P. G. Tratnyek. 1994. Reductive dechlorination of chlorinated methanes by iron metal. Environmental Science & Technology 28:2045-2053.

Maxymillian, N. A., S. A. Warren, and E. F. Neuhauser. 1994. Thermal Desorption of Coal Tar Contaminated Soils from Manufactured Gas Plants. Pittsfield, Mass.: Maxymillian Technologies.

McCarty, P. L. 1996. An overview of anaerobic transformation of chlorinated solvents. Paper presented at Conference on Intrinsic Remediation of Chlorinated Solvents, Airport Hilton, Salt Lake City, Utah, April 2, 1996.

McCarty, P. L., and L. Semprini. 1994. Ground water treatment for chlorinated solvents. Section 5 in Handbook of Bioremediation, R. D. Norris, ed. Boca Raton, Fla.: CRC Press.

Means, J. C., S. G. Wood, J. J. Hassett, and W. L. Banwart. 1980. Sorption of polynuclear aromatic hydrocarbons by sediments and soils. Environmental Science & Technology 14:1524-1528.

Mercer, J. W., and R. M. Cohen. 1990. A review of immiscible fluids in the subsurface: Properties, models, characterization and remediation. Journal of Contaminant Hydrology 6:(2)107-163.

Miller, R. N., R. E. Hinchee, C. M Vogel, R. R. DuPont, and D. C. Downey. 1990. Field investigation of enhanced petroleum hydrocarbon degradation in the vadose zone of Tyndall AFB. In Proceedings, Petroleum Hydrocarbons and Organic Chemicals in Groundwater: Prevention, Detection, and Remediation, Houston, October 31-November 2, 1990. Worthington, Ohio: National Water Well Association.

Mihelcic, J. R. and R. G. Luthy. 1988. Degradation of polycyclic aromatic hydrocarbon compounds under various redox conditions in soil-water systems. Applied and Environmental Microbiology 54:1182-1187.

Mohn, W. W., and J. M. Tiedje. 1992. Microbial reductive dehalogenation. Microbial Review 56:482-507.

Morea, S. C., R. L. Olsen, and R. W. Chappelle, in conjunction with scientists and engineers from the Colorado School of Mines and Denver Knight Piesold. 1989. Assessment of a passive treatment system for acid mine drainage. Paper presented at 62nd Annual Conference of the Water Pollution Control Federation, San Francisco, California, October 16-19, 1989.

Morgan, D., A. Battaglia, B. Hall, L. Vernieri, and M. Cushney. 1992. The GRI Accelerated Biotreatability Protocol for Assessing Conventional Biological Treatment of Soils: Development and Evaluation Using Soils from Manufactured Gas Plant Sites. Technical Report GRI-92/0499. Chicago, Ill.: Gas Research Institute.

Nakles, D., D. Linz, and I. Murarka. 1991. Bioremediation of MGP soils: Limitations and potentials. Presented at EPRI Technology Transfer Seminar on Management of Manufactured Gas Plant Sites, Atlanta, Ga., April 2-3, 1991.

National Research Council. 1993. In Situ Bioremediation: When Does It Work? Washington, D.C.: National Academy Press.

National Research Council. 1994. Alternatives for Groundwater Cleanup. Washington, D.C.: National Academy Press.

Norris, R. D., and J. E. Matthews. 1994. Handbook of Bioremediation. Ann Arbor, Mich.: Lewis Publishers.

Office of Naval Research, Air Force Office of Scientific Research, Army Research Office, and 7Army Corps of Engineers, Waterways Experiment Station. 1995. A Tri-Service Workshop on Bioavailability of Organic Contaminants in Soils and Sediments, Monterey, Calif., April 9-12, 1995. Arlington, Va.: Office of Naval Research.

Oliver, B. G. 1985. Desorption of chlorinated hydrocarbons from spiked and anthropogenically contaminated sediments. Chemosphere 14(8):1087-1106.

Ong, S. K., A. Leeson, R. E. Hinchee, J. Kittel, C. M. Vogel, G. D. Sayles, and R. N. Miller. 1994. Cold climate applications of bioventing. Pp. 444-453 in Hydrocarbon Bioremediation, R. E. Hinchee, B. C. Alleman, R. E. M. Hoeppel, and R. N. Miller, eds. Ann Arbor, Mich.: Lewis Publishers.

Palmer, C. D., and P. R. Wittbrodt. 1991. Processes affecting the remediation of chromium-contaminated sites. Environmental Health Perspectives 92:25-40.

Pankow, J. F., and J. A. Cherry, eds. 1996. Dense Chlorinated Solvents and Other DNAPLs in Ground Water: History, Behavior, and Remediation. Portland, Oreg.: Waterloo Press.

Pennell, K. D., G. A. Pope, and L. M. Abriola. 1996. Influence of viscous and buoyancy forces on the mobilization of residual tetrachloroethylene during surfactant flushing. Environmental Science & Technology 30(4):1328-1335.

Pope, G. A., and W. H. Wade. 1995. Lessons learned from enhanced oil recovery research for surfactant enhanced aquifer remediation. In Surfactant Enhanced Subsurface Remediation: Emerging Technologies. Washington, D.C.: American Chemical Society.

Powell, R. M., R. W. Puls, S. K. Hightower, and D. A. Sabatini. 1995. Coupled iron corrosion and chromate reduction: Mechanisms for subsurface remediation. Environmental Science & Technology 29:1913-1922.

Puls, R. W., R. M. Powell, and C. J. Paul. 1995. In situ remediation of ground water contaminated with chromate and chlorinated solvents using zero-valent iron: A field study. Paper presented at the 209th American Chemical Society National Meeting, Anaheim, Calif., April 2-7, 1995. Washington, D.C.: American Chemical Society, Division of Environmental Chemistry.

Raymond, R. L., V. W. Jamison, and J. O. Hudson. 1977. Beneficial stimulation of bacterial activity in groundwater containing petroleum hydrocarbons. American Institute of Chemical Engineers Symposium Series 73(166):390-404.

Rice, D., B. P. Dooher, S. J. Cullen, L. G. Everett, W. E. Kastenberg, R. D. Gose, and M. A. Marino. 1995. California Leaking Underground Fuel Tank (LUFT) Historical Case Analyses. UCRL-AR-122207. Livermore, Calif.: Lawrence Livermore National Laboratory.

Roberts, A. L., L. A. Totten, W. A. Arnold, D. R. Burris, and T. J. Campbell. 1996. Reductive elimination of chlorinated ethylenes by zero-valent metals. Environmental Science & Technology 30:2654-2659.

Roy, K. 1989. Electrifying soil cleanup. HazMat World (July):14-15.

Roy, S. B., D. A. Dzombak, and M. A. Ali. 1995. Assessment of in situ solvent extraction for remediation of coal tar sites: Column studies. Water Environment Research 67(1):4-15.

Salanitro, J. 1993. An industry's perspective on intrinsic bioremediation. Pp. 104-109 in In Situ Bioremediation: When Does It Work? Washington, D.C.: National Academy Press.

Schwille, F. 1988. Dense Chlorinated Solvents in Porous and Fractured Media (translated by J. F. Pankow). Chelsea, Mich.: Lewis Publishers.

Semprini, L., P. V. Roberts, G. D. Hopkins, and P. L. McCarty. 1990. A field evaluation of in-situ biodegradation of chlorinated ethenes: Part 2—the results of biostimulation and biotransformation experiments. Groundwater 28:715-727.

Sheldon, R. B. 1996. Field demonstration of a full scale in situ thermal desorption system for the remediation of soil containing PCBs and other hydrocarbons. Presented at Superfund XVII, Washington, D.C., October 1996.

Singhvi, R., R. N. Koustas, and M. Mohn. 1994. Contaminants and Remedial Options at Pesticide Sites. EPA/600/R-94/202. Washington, D.C.: EPA.

Smith, J., P. Tomiceck, R. Weightman, D. Nakles, D. Linz, and M. Helbling. 1994. Definition of biodegradation endpoints for PAH contaminated soils using a risk-based approach. Paper presented at the Ninth Annual Conference on Contaminated Soils Using a Risk-Based Approach, University of Massachusetts, Amherst, Mass., October 18-20, 1994.

Smith, J. R., M. E. Egbe, and W. J. Lyman. In press. Bioremediation of polycholorinated biphenyls (PCBs) and polynuclear aromatic hydrocarbons (PAHs). In Bioremediation of Contaminated Soils. Madison, Wis.: American Society of Agronomy/Soil Science Society of America.

Smith, L. A., B. C. Alleman, and L. Copley-Graves. 1994. Biological treatment options. Pp. 1-12 in Emerging Technology for Bioremediation of Metals, J. L. Means and R. E. Hinchee, eds. Ann Arbor, Mich.: Lewis Publishers.

Smith, L. A., J. L. Means, A. Chen, B. Alleman, C. C. Chapman, J. S. Tixier, Jr., S. E. Brauning, A. R. Gavaskar, and M. D. Royer. 1995. Remedial Options for Metals-Contaminated Sites. Boca Raton, Fla.: Lewis Publishers/CRC Press.

Soundararajan, R. 1992. Guidelines for evaluation of the permanence of a stabilization/solidification technology. Pp. 33-39 in Stabilization and Solidification of Hazardous, Radioactive, and Mixed Wastes, Vol. 2, T. M. Gilliam and C. C. Wiles, eds. ASTM STP 1123. Philadelphia: American Society for Testing and Materials.

Stinson, M. K. 1990. EPA SITE demonstration of the international waste technologies/geo-con in situ stabilization/solidification process. Journal of Air and Waste Management Association 40:1569-1576.

Stinson, M. K., H. Skovronek, and W. D. Ellis. 1992. EPA SITE demonstration of the BioTrol soil washing process. Journal of Air and Waste Management Association 42:96-103.

Swiss Federal Institute for Environmental Science and Technology. 1994. Workshop on biological degradation of polycyclic aromatic hydrocarbons (PAHs) in soil: state of the art and identification of research needs, Zurich, Switzerland, April 7-8, 1994.

Texas Research Institute. 1982. Enhancing the Microbial Degradation of Underground Gasoline by Increasing Available Oxygen, Final Report. Washington, D.C.: American Petroleum Institute.

Thomson, B. M., and W. R. Turney. 1995. Minerals and mine drainage. Water Environment Research 67:527-529.

Thornton, J. C., and W. L. Wooten. 1982. Venting for the removal of hydrocarbon vapors from gasoline contaminated soil. Journal of Environmental Science and Health AI7(1):31-44.

Treigel, E. K., and L. Guo. 1994. Overview of the fate of pesticides in the environment, water balance; runoff vs leaching. Pp. 1-13 in Mechanisms of Pesticide Movement Into Ground Water. R. C. Honeycutt and D. J. Schabacker, eds. Boca Raton, Fla.: Lewis Publishers.

Trobridge, T. D., and T. C. Halcombe. 1994. Waste treatment via solvent extraction/dehydration with the Carver-Greenfield process. Presented at the I&EC Special Symposium, Atlanta, September 19-21, 1994. Washington, D.C.: American Chemical Society.

Troxler, W. L., J. J. Cudahy, R. P. Zink, S. I. Rosenthal, and J. J. Yezzi. 1992. Treatment of petroleum contaminated soils by thermal desorption technologies. Presented at 85th Annual Meeting of the Air and Waste Management Association, Kansas City, Mo., June 21-26, 1992.

Udell, K. S., and L. D. Stewart. 1989. Field Study of In Situ Steam Injection and Vacuum Extraction for Recovery of Volatile Organic Solvents. UCB-SEEHRL Report No. 89-2. Berkeley, Calif.: University of California, Environmental Health Research Laboratory.

Udell, K. S., and L. D. Stewart. 1990. Combined steam injection and vacuum extraction for aquifer cleanup. Presented at the International Association of Hydrologists Conference, Calgary, Alberta, Canada, April 1990.

van der Leeden, F., F. L. Troise, and D. K. Todd, eds. 1990. The Water Encyclopedia. Chelsea, Mich.: Lewis Publishers.

Vidic, R. D., and F. G. Pohland. 1996. Treatment Walls. Technology Evaluation Report TE-96-01. Pittsburgh, Pa.: Ground Water Remediation Technologies Analysis Center.

Vorum, M. 1991. Dechlorination of polychlorinated biphenyls using the SoilTech anaerobic thermal processing unit, Wide Beach Superfund site, New York. Presented at HazTech International 91, Pittsburgh, Pa., May 14-16, 1991.

Watanabe, M. 1997. Phytoremediation on the brink of commercialization. Environmental Science & Technology 31(4):182A-186A.

Weitzman, L., and L. E. Howel. 1989. Evaluation of Solidification/Stabilization as a Best Demonstrated Available Technology for Contaminated Soils. EPA/600/2-89/049. Washington, D.C.: EPA.

Weitzman, L., K. Gray, F. K. Kawahara, R. W. Peters, J. Verbicky. 1994. Chemical Treatment. Vol. 2 of Innovative Site Remediation Technology, W. C. Anderson, ed. Annapolis, Md.: American Academy of Environmental Engineers.

Whiting, K., R. L. Olsen, J. N. Cevaal, and R. Brown. 1994. Treatment of mine drainage using a passive biological system: Comparison of full-scale results to bench- and pilot-scale results. In Proceedings of the Society for Mining, Metallurgy and Exploration Annual Conference, Albuquerque, New Mexico, February 14, 1994. Littleton, Colo.: Society for Mining, Metallurgy and Exploration.

Wiedemeier, T. H., J. T. Wilson, D. H. Campbell, R. Miller, and J. Hansen. 1995. Technical Protocol for Implementing Intrinsic Remediation with Long Term Monitoring for Natural Attenuation of Fuel Contamination in Ground Water. Brooks Air Force Base, Tex.: Air Force Center for Environmental Excellence.

Wiedemeier, T. H., L. A. Benson, J. T. Wilson, D. H. Kampbell, and R. Miknis. 1996. Patterns of natural attenuation of chlorinated aliphatic hydrocarbons at Plattsburgh Air Force Base, New York. Presented at Conference on Intrinsic Remediation of Chlorinated Solvents, Airport Hilton, Salt Lake City, Utah, April 2, 1996.

Wildeman, T. R. 1992. Constructed wetlands that emphasize sulfate reduction. Paper 32 in Proceedings of the 24th Annual Operators Conference of the Canadian Mineral Processors. Ottawa, Ontario: Canadian Institute of Mining, Metallurgy, and Petroleum.

Wildeman, T. R., S. P. Machemer, R. W. Klusman, R. H. Cohen, and P. Lemke. 1990. Metal removal efficiencies from acid mine drainage in the Big Five constructed wetland. Pp. 417-424 in Proceedings of the Mining and Reclamation Conference and Exhibition, Charleston, West Virginia, April 23-26, 1990, J. Skousden, J. Sencindiver, and D. Samuel, eds. Washington, D.C.: U.S. Bureau of Mines.

Wildeman, T. R., D. M. Updegraff, J. S. Reynolds, and J. L. Bolis. 1994. Passive bioremediation of metals from water using reactors or constructed wetlands. Pp. 13-25 in Emerging Technology for Bioremediation of Metals, J. L. Means, and R. E. Hinchee, eds.. Ann Arbor: Lewis Publishers.

Wiles, C. C., and E. Barth. 1992. Solidification/stabilization: Is it always appropriate? Pp. 18-32. in Stabilization and Solidification of Hazardous, Radioactive, and Mixed Wastes, Vol. 2, T. M. Gilliam, and C. C. Wiles, eds. ASTM STP 1123. Philadelphia: American Society for Testing and Materials.

Wilson, E. K. 1995. Zero-valent metals provide possible solution to groundwater problems. Chemical & Engineering News (July 3):19-22.

Wilson, J. T., D. H. Kampbell, and J. Armstrong. 1994. Natural bioreclamation of alkylbenzenes (BTEX) from a gasoline spill in methanogenic groundwater. Pp. 201-218 in Hydrocarbon Bioremediation, R. E. Hinchee, B. C. Alleman, R. E. M. Hoeppel, and R. N. Miller, eds. Ann Arbor, Mich.: Lewis Publishers.

Woodruff, R. K., R. W. Hanf, and R. E. Lundgren, eds. 1993. Hanford Site Report for Calendar Year 1992. PNL-86821UC-602. Richland, Wash.: Pacific Northwest Laboratory.

Yamane, C. L., S. D. Warner, J. D. Gallinatti, F. S. Szerdy, T. A. Delfino, D. A. Hankins, and J. L. Vogan. 1995. Installation of a subsurface groundwater treatment wall composed of granular zero-valent iron. Pp. 792-795 in Preprints of Papers Presented at the 209th American Chemical Society National Meeting, Anaheim, Calif., April 2-7, 1995. Washington, D.C.: American Chemical Society, Division of Environmental Chemistry.

4

Measures of Success for Remediation Technologies

Development and implementation of innovative technologies for ground water cleanup is shaped by many diverse and sometimes contradictory expectations of what constitutes success. While many industries, such as the automotive and aerospace industries, have developed uniform standards for evaluating product performance, no such standards exist for ground water and soil remediation technologies. Property owners responsible for site cleanup, citizen groups, state and federal regulators, and technology developers all may have different perspectives on how remediation technologies should be evaluated and selected. Reconciling the differing expectations of these stakeholders can add to delays in site remediation. A standard approach for comparing remediation technology performance could lead to a less contentious (and less time-consuming) technology selection process and possibly to improved acceptance of innovative remediation technologies. The challenge is to relate the success criteria important to the many stakeholders to specific technology performance criteria that can be measured or at least accounted for in some uniform way.

This chapter provides an overview of the many criteria that can be important to different stakeholder groups when evaluating ground water and soil cleanup technologies. The success criteria can be divided into three categories: (1) technological performance, (2) commercial characteristics, and (3) acceptability to the public and regulators. Table 4-1 lists the key success criteria, which are discussed in detail later in this chapter, in each of these categories. The rankings of high (H), medium (M), or low (L) interest reflect the committee's assessment of the average level of importance of each criterion to each stakeholder group. Technical performance attributes describe the technology's ability to achieve risk reduction goals and the efficiency with which it achieves these goals. Commercial

TABLE 4-1 Stakeholders' Concerns About Remediation Technology Performance

Technology Performance Attribute	Stakeholders						
	Public	Regulators	Technology Users	Technology Providers	Investors	Insurance Companies	Site Workers
Technical performance							
Reduction in health and environmental risk (contaminant mass, concentration, toxicity, and mobility)	H	H	H	H		H	
Robustness[a]		M	H	H			
Forgiveness[b]		L	H	H			
Ease of implementation		L-M	H	H	M		
Maintenance; down time		M	H	H	H		
Predictability; ease of scaleup	M	H	H	H	M		
Secondary emissions and residuals production	H	H	M	M		M	H
Commercial characteristics							
Capital costs	L	M-H	H	M	H	M	
Operating costs	L	M-H	H	M	L		M
Copyright, patent restrictions				H	H		
Profitability				H	H		
Accessibility	L-M	M-H	M-H	H	H		

Acceptability to the public and regulators						
Disruption to community	H	M			M	H
Disruption to ongoing site activity	M	H			H	
Safety	H	M	H		H	H
Regulatory hurdles	H	M	H	H	H	H
Future usability of the land	H	M-H	H		L	H

NOTE: "H" denotes a high, "M" denotes a medium, and "L" denotes a low level of interest in the indicated technology performance attribute. A blank entry denotes no or very minimal concern about the particular performance attribute.

[a]Robustness refers to a technology's ability to operate over a range of environmental conditions.
[b]Forgiveness refers to a technology's sensitivity to operating conditions.

characteristics are factors related to the costs of the technology and the profits it yields. Public and regulatory acceptance attributes are qualitative characteristics of technology performance that, in addition to quantitative technology attributes, are of particular importance to the public near the contaminated site and to regulators; to varying degrees, these attributes also may be important to other stakeholder groups. Chapter 5 provides details about how to evaluate technical performance of technologies. Chapter 6 describes how to assess commercial characteristics of technologies. Not all public and regulatory acceptance attributes can be measured quantitatively, but they nevertheless must be considered when developing a new technology.

To be successful, a technology must have strengths in all three of the areas shown in Table 4-1: technical performance, commercial viability, and appeal to the public and regulators. One remediation technology that illustrates success in meeting these three categories of criteria is soil vapor extraction (see Box 4-1).

STAKEHOLDER CRITERIA FOR SUCCESS

To be widely applied, a remediation technology must be not only a success in that it meets technical performance criteria, but it also must be accepted by numerous stakeholders who have an interest in the application of the technology. Expectations about how a technology should perform can vary widely among the key stakeholder groups: the public, regulators, technology users or consumers, investors in innovative technology, insurance companies, and individuals working at the site.

The Public

The key members of the public to consider when evaluating technology performance are those living near the contaminated site. Active local communities can and often do block implementation of remediation technologies that they perceive as unacceptable. The most important aspect of remediation technology performance for communities near contaminated sites is usually the degree to which the technology can reduce risks to community health and the local environment. For example, residents of Woburn, Massachusetts, became active in calling for site remediation because they believed there was an association between a cluster of childhood leukemia incidences and the contamination of two town wells with industrial solvents (Brown and Mikkelson, 1990). Along the Housatonic River in western Massachusetts, citizens have worked for years to address polychlorinated biphenyl (PCB) contamination of sediments because of the desire to maintain the river's value for recreation and harvesting of aquatic species (Ewusi-Wilson et al., 1995).

The cost of remediation may be a concern of the public at large, but in a community with a contaminated site (where remediation costs are rarely experi-

---

BOX 4-1
Soil Vapor Extraction (SVE): Technology Success Story

SVE is an example of a remediation technology that progressed rapidly through the development and commercialization process because it met many of the criteria for success listed in Table 4-1. SVE systems have been selected for use at Superfund sites more than any other type of innovative technology (McCoy and Associates, 1993; EPA, 1996). They are in use at thousands of other contaminated sites, especially sites contaminated with gasoline from leaking underground storage tanks. Following are some of the key attributes of SVE technology that have led to its success:

• *Technical performance:* In addition to a well-documented ability to reduce contaminant mass and concentration, SVE is easy to engineer. It is robust over a range of contaminant conditions; essentially, it works for any volatile compound. It is easy to design, and there are now many examples that can be used as the basis for future designs. It requires no sophisticated equipment or operators, so operation and maintenance are relatively easy.

• *Commercial characteristics:* Because SVE requires no excavation of contaminated soil, capital costs are low. The simplicity of operation makes maintenance costs relatively low, as well. SVE is easily accessible because of the large number of engineering firms offering SVE design services. This accessibility has kept the price low for the end user due to widespread competition.

• *Acceptablity to the public and regulators:* Because SVE requires no soil excavation, community disruption is minimal. The technology is safe to operate. In situations where contaminant vapors have entered buildings, the ability of SVE to address these vapors is rapid and obvious, leading to a public perception of its benefit. SVE now has a track record of success in being approved by regulators and in achieving the regulatory goals required for site closure.

---

enced directly), there may well be an interest in identifying the most effective technology, regardless of cost. Members of the public who believe they have experienced health damage or who believe extensive natural resource damage has occurred may resent efforts by government agencies or responsible parties to minimize remediation costs.

Other factors important to local communities include the safety of the technology and the degree to which it will disrupt the community. For example, selecting thermal destruction or desorption may lead to air emissions of toxic byproducts (such as dioxins or furans), adversely affecting the neighborhoods near the site. Similarly, in situ flushing technologies may result in uncontrolled migration of contaminants to previously uncontaminated zones. Excavating contaminated soil and storing it on site while awaiting treatment can generate dust, which can increase exposure risks, at least for short periods. When there is a personal impact, such as digging up yards, encroaching on property, or creating excess truck traffic, the affected community members will carefully weigh these impacts against the environmental benefit of cleaning up the site. Because innovative technologies, by definition, are not used at many sites, public reluctance to accept an unproven technology may be as great as that of site owners or regulators.

Compounding the challenge of gaining acceptance of a technology is the public's realization that experts often disagree about the nature of risks, the degree to which a site should or can be cleaned up, and the effectiveness of a particular technology (Kraus et al., 1992). This lack of certainty, coupled with the fact that the solutions are being selected by nonresidents (regulatory agencies and those responsible for the contamination) often fosters distrust, not only of the decisionmakers but also of the technologies they propose. Community members are more likely to accept a remediation technology if they have been active participants in the investigation and remediation process. Because local residents often have historical knowledge of the community, they may offer valuable input during the early stages of site investigation.

Regulators

Regulatory agencies seek proof that a remediation technology can meet the requirements of the various statutes governing site cleanup, including the Comprehensive Environmental Response, Compensation, and Liability Act (Superfund), the Resource Conservation and Recovery Act (RCRA), and the state-level equivalents of these two programs (see Chapter 1). These statutes are based on protection of human health and the environment, and the associated regulatory criteria are generally health based. Cost is also a concern for regulatory agencies, especially those at the state level. If responsible parties are local industries, extraordinarily high remediation costs may result in a threat to shut down operations and move out of state, resulting in loss of jobs and tax base. On the other hand, if revenues come from a state fund, high expenditures on one project may mean fewer dollars for others or, alternatively, may mean going back to reluctant sources to replenish the fund. This latter scenario has slowed progress in cleaning up leaking underground storage tanks; many of the state trust funds established to facilitate these cleanups ran dry before the cleanups could be completed.

Regulatory agencies are subject to substantial public scrutiny with respect to the efficacy of the technology selected, its cost, and the successful implementation of the technology, including holding to original cost projections for installation and operation. Community disruption is a concern for regulators because community discontent often is directed at the agency. Consequently, regulatory agencies are interested in using remediation technologies that operate effectively over a range of conditions and are safe, because these have a lesser chance of producing embarrassing incidents.

Technology Providers

Technology developers and owners undertake their efforts in part because they are interested in solving complex problems that have real-world applications, in part because they hope to profit, and in part from a conviction that the approach they have conceived can achieve objectives better, faster, or cheaper than the conventional technology. Sources of innovative technology vary. Basic science research conducted in the academic community and in government laboratories is the source of many innovative remediation technologies. Innovative technologies also have been developed by technology service providers such as engineering and consulting firms and by companies responsible for site cleanup, who then become technology users.

The market for innovative remediation technologies is highly segmented and profoundly influenced by laws and regulations. Laws and regulations create the impetus for agencies and private parties to undertake remedial actions, and, with some exceptions, government approvals are usually needed for the choice of remediation technology.[1] Thus, technology owners have a significant interest in showing that their technology is effective in meeting the government's requirements. They also have a financial interest in promoting widespread use of the technology, which means demonstrating that it is applicable under a range of site conditions and competitive with other technologies that address the same needs.

Technology Users

Technology users (clients) are, in many cases, reluctant customers. As discussed in Chapter 2, those responsible for contamination may take a variety of actions to defer using any remediation technology. When all other alternatives have been exhausted and a technology choice must be made, remediation technology users usually focus primarily on meeting regulatory standards for risk reduction as cost effectively and expeditiously as possible. The technology user has a strong stake in the remediation technology working right the first time and is concerned about the technology's ease of implementation, robustness over a

---

[1]For voluntary cleanups, cleanups occurring as part of a RCRA interim measure, and some cleanups occurring under state hazardous waste programs, government approval of the remediation technology may not be necessary.

range of site conditions, ability to handle variable waste streams, interference with ongoing activities at the site, and maintenance requirements.

Investors

Investors are most concerned about factors related to a technology's potential to generate profits. Venture capitalists, like technology owners, may be concerned about whether the technology can meet regulatory requirements. As explained in Chapter 2, at a more fundamental level, they may be concerned about whether the regulatory environment is sufficiently predictable that a market exists for the technology. A widely held perception among venture capitalists is that the present regulatory system creates incentives for responsible parties to delay making investments in technology, thus postponing until some future point the demand for widespread use of cleanup technologies (see Chapter 2).

Once a clear market exists for a technology, investors will be concerned about its affordability to potential customers and whether it is applicable to a wide range of site conditions. Specialized technologies with very limited applications may not be sufficiently profitable to attract investor interest. Investors may also be concerned about patents or copyrights. Innovations that are protected by patents or copyrights are attractive to investors because of the potential for licensing agreements, royalties, and other arrangements that may generate a significant income stream.

Insurance Companies

Insurance companies, when involved in site remediation, are most concerned about minimizing current and future liability. Factors related to current liability include the creation of residuals, safety (both on and off the site), and disruption to the community. To minimize long-term liability, insurance companies will want proof that the technology can reduce or eliminate health and environmental risks so that they will not face continuing financial liabilities. Insurance companies paying for site remediation will also be concerned about costs of the remediation technology.

Site Workers

Individuals working at the remediation site may include those involved in implementing the remediation technology and those who work at a site at which remediation is occurring but are employed in other activities. An example of the second situation is a RCRA corrective action site at which cleanup is taking place at an existing manufacturing operation. In this case, workers will continue to perform their manufacturing activities. This second group of site workers, many of whom may live in the local community, may be concerned about whether cleanup costs will be so high that they will lose their jobs. Both groups of site workers may be concerned about safety and risks to their health.

Roles of Stakeholders in the Site Cleanup Process

Contaminated sites became an issue for stakeholders in response to public outrage and media attention to environmental damage and suspected human health effects. At both the state and federal levels, regulatory agencies have established processes for making decisions about which sites are sufficiently contaminated that they must be cleaned up and for selecting cleanup technologies for those sites. Figure 4-1 is a flow chart representing the process used for decisions about remediation at Superfund sites. Similar processes are in place in many state-level site remediation programs.

The decision process is somewhat different for RCRA corrective action sites because ongoing operations take place at these sites. The primary difference is the absence of the national ranking and listing step that takes place under Superfund. In addition, interactions between the site owner and regulators during the remediation technology selection process are generally less contentious under RCRA than under Superfund. Instead of being specified in a legally negotiated record of decision (ROD), remediation technology selections under RCRA are included in the overall permit for treating, storing, and disposing of hazardous wastes as part of ongoing operations at the site.

Common to all programs like Superfund is a relatively complex and very time-consuming chain of events in which studies are performed and evaluated primarily by agencies and site owners. Although this lengthy process may meet the need of the regulators for order and accountability, it also often delays remedial activities. For example, at the site described in Box 4-2, final remedies have not yet been installed, although site investigations began in 1975.

In practice, the process of technology selection may be less linear than shown in Figure 4-1. For example, in the case of the Pine Street Canal (Box 4-2), an additional remedial investigation and feasibility study was undertaken in response to the public's overwhelming negative reaction to the Environmental Protection Agency's (EPA's) proposed remedy. In the case of the Caldwell Trucking Superfund site (see Box 4-3), the EPA prepared a ROD following the public hearing, but public opposition to the ROD was sufficiently vehement that a modification was sought and approved. In general, the regulatory agency expects all of the parties responsible for contamination to implement whatever actions are embodied in the ROD it prepares. The regulatory agency commitment to a linear process can be a problem in itself; a linear progression constrains flexibility in exploration and may preclude acting effectively on data that are generated late in the process.

Stakeholders enter into the evaluation process at different times and have varying degrees of influence on selection of the final remedy. Table 4-2 lists the

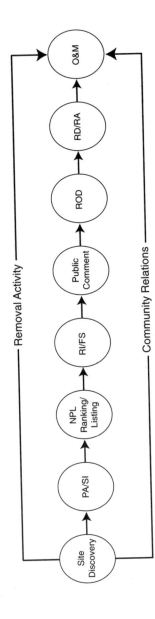

SITE DISCOVERY: A site is identified and reported to EPA, and the Superfund process begins.

PA-PRELIMINARY ASSESSMENT: The process of collecting and reviewing available information about a known or suspected hazardous waste site or release. EPA or states use this information to determine if the site requires further study. If further study is needed, a site inspection (SI) is undertaken.

SI-SITE INSPECTION: A technical phase that follows a preliminary assessment designed to collect more extensive information on a hazardous waste site. The information is used to score the site with the Hazard Ranking System (HRS) to determine whether response action is needed.

NPL RANKING/LISTING- NATIONAL PRIORITIES LIST RANKING/LISTING: A site is ranked using the HRS, a process that evaluates site conditions and determines whether a site should be proposed for the National Priorities List (NPL). The NPL is EPA's list of the most serious uncontrolled or abandoned hazardous waste sites identified for possible long-term remedial response. EPA is required to update the NPL at least once a year.

177

RI/FS-REMEDIAL INVESTIGATION/FEASIBILITY STUDY: Two distinct studies carried out concurrently. The RI/FS is usually performed as the first phase of the Superfund remedial process for sites on the NPL. The RI/FS is intended to (1) gather the data necessary to determine the type and extent of contamination at a Superfund site; (2) establish criteria for cleaning up the site; (3) identify and screen cleanup alternatives for remedial action; and (4) analyze in detail the technology and costs of the alternatives at the Superfund site.

PUBLIC COMMENT: An opportunity for the site community and interested parties to ask questions about and comment on the remedial activities.

ROD-RECORD OF DECISION: A public document that explains which cleanup alternative(s) will be used at an NPL site. The ROD is based on information and technical analysis generated during the RI/FS, with consideration of public comments and community concerns.

RD-REMEDIAL DESIGN: An engineering phase of the Superfund process that follows the ROD; technical drawings and specifications are developed for the subsequent remedial action at a site on the NPL.

RA-REMEDIAL ACTION: The actual construction or implementation phase of the Superfund process that follows the remedial design of the selected cleanup alternative at a site on the NPL.

O&M-OPERATION AND MAINTENANCE: Activities conducted at a site once a response action is concluded to ensure that the cleanup or containment system is functioning properly.

REMOVAL ACTIVITIES: Intermediate Superfund actions taken over the short term to address a release or threatened release of hazardous substances.

COMMUNITY RELATIONS: EPA's program to inform and involve the public in the Superfund process and respond to community concern.

FIGURE 4-1   Steps in the Superfund remedial process.

BOX 4-2
Pine Street Canal: Public Concern Leads to the Search for Innovative Technologies

Placed on the Superfund National Priorities List in 1981, the 70-acre Pine Street Canal Superfund site on the eastern shore of Lake Champlain in Burlington, Vermont, is the focus of considerable community interest. The site is bordered to the west by mixed commercial and residential neighbors and contains wetlands, surface water, and upland areas, some of which were created by fill. To date, no remediation technology has been selected.

The Pine Street Canal has been used for a variety of commercial and industrial purposes from the 1800s to the present. In the mid-1800s, a barge canal and turning basin were built to serve sawmills, lumber yards, a boat yard, and several coal yards. The site also hosted a manufactured gas plant that produced heating, cooking, and illuminating gas for the Burlington area from the late 1800s until 1966 (GEI Consultants, Inc., 1995). Other former or current operations include brush fiber manufacturing, helicopter manufacturing, magnesium casting, metal finishing, petroleum storage, and asphalt batching. The coal gasification plant has been identified as the primary source of contamination, but contamination likely resulted from all of the other uses as well (GEI Consultants, Inc., 1995). Site investigations have shown the primary contaminants to be benzene, toluene, ethylbenzene, and xylenes (BTEX); polycyclic aromatic hydrocarbons (PAHs); and metals (GEI Consultants, Inc., 1995).

In 1992, the Environmental Protection Agency's (EPA's) proposed plan for the site called for dredging contaminated canal sediments, excavating contaminated surface soils, and placing the contaminated materials in an on-site confined disposal facility. Long-term hydraulic controls

---

primary decision points in the Superfund process and indicates, for each stakeholder, the degree of participation at each decision point. The term "technology users" refers to stakeholders who are purchasing technology; in Table 4-2, that definition is expanded to include the subcontractors and consultants hired by the user to help generate data and recommendations as part of the decisionmaking process. In the language of Superfund, the technology users are called potentially responsible parties (PRPs), or responsible parties.

Table 4-2 shows that, except for the initial comment period, the level of public participation in the remedy selection process is generally low, in large part because there is no routine mechanism for public involvement other than the comment period. Participation of the technology providers depends on particulars of the site, but the point at which they have high participation is in the remedial design and remedial action stage. Both regulators and technology users have an

were proposed for the disposal facility. Community opposition to the EPA proposal focused on the large volume of material to be excavated (R. L. Gilleland, EPA Region I, personal communication, 1995) and on the placement of the confined disposal facility in an on-site wetland. EPA withdrew its proposal in June 1993. The Pine Street Canal Coordinating Council was convened at that time and charged with recommending to the EPA a remedial plan for the site (GEI Consultants, Inc., 1995). Because this stakeholder involvement occurred after extensive work by the EPA, it was necessary to go back in the remedial process and repeat some steps.

Additional remedial investigation and feasibility study work has been carried out under a consent order. The public comments on the EPA's 1992 proposal resulted in suggestions for alternative remedial action technologies that are being assessed as part of the additional work; possible use of innovative technology was initially suggested in the public comment period. The statement of work requires the evaluation of various remedial technologies including those in the original proposal and "other non-intrusive engineered techniques, as well as innovative treatment technologies used at existing Manufacturing Gas Plant (MGP) sites and recycle/reuse options" (EPA Region I, 1995; O'Donnell, 1995).

Members of the Pine Street Coordinating Council have a shared interest in innovative technology. According to Lori Fisher, executive director of the Lake Champlain Committee and Coordinating Council member, each of the stakeholder groups on the council, the potentially responsible parties, the EPA, the state government, and the various community interests can see advantages to innovative technology. Fisher notes, however, that the site characteristics and the contaminants (BTEX, PAHs, and metals, which are commingled in some locations) pose significant challenges (L. Fisher, personal communication, 1996).

interest in all stages of the process, but technology users have minimal participation in the National Priorities List ranking (although they can appeal the ranking), which is a government activity.

Both the Pine Street Canal (Box 4-2) and the New Bedford Harbor (Box 4-4) cases are situations in which public opposition to EPA recommendations for using conventional technology has resulted in modifications to the decisionmaking process. When the decisionmaking process was altered and some steps were revisited, innovative technologies were given serious consideration. These cases suggest that if the public were involved earlier in the decisionmaking process as a matter of routine, the universe of remediation technologies considered at sites might more routinely include innovative technologies.

In summary, different stakeholders may have quite different concerns about the selection and use of a remediation technology at a given contaminated site.

BOX 4-3
The Triumph and Caldwell Trucking Superfund Sites:
Communities Reject Aggressive Cleanup Remedies

In the town of Triumph, Idaho (an old mining town of 50 residents near Sun Valley), residents have staunchly opposed a cleanup scheduled to occur under the federal Superfund program (Stuebner, 1993; Gallagher, 1993; Miller, 1995). The cleanup could affect the property of about a dozen homeowners. It would involve removal of soil contaminated with arsenic and lead tailings.

The cleanup level is based on total soil metals content. The community believes that this required cleanup level is inappropriate because community blood-lead and arsenic-urine tests revealed no acute health problems from heavy metals. Blood-lead levels are below the national average. The citizens believe the metals pose no threat because they are not bioavailable and that soil removal would provide no improvement. Consequently, there is continuing opposition to any mandated cleanup under Superfund. In this case the issue goes beyond selection of a cleanup technology to the point that the community wants no action whatsoever.

At the Caldwell Trucking Company Superfund site in Fairfield Township, New Jersey, the community rejected a pump-and-treat remedy to slowly decontaminate a plume of contaminated ground water in favor of a less expensive hydraulic containment system (EPA Region II, 1993). Industrial and septic wastes had been disposed of at the site, releasing significant amounts of tricholoethylene in both a shallow water table aquifer and bedrock aquifer. Migration was toward the Deepavaal Brook and Passaic River about a mile to the north of the site. The ground water is not used for drinking water. The EPA preferred remedy in 1989 was to pump and treat the entire aquifer system for more than 30 years to reduce contaminant concentrations to acceptable levels, recognizing that fully cleaning up the aquifers would take more than 100 years. This system would have required an extensive network of wells and piping to be maintained within the surrounding community.

Virtually all of the community's 100 members were opposed to the disruption that the pump-and-treat network would bring. Consequently, a modification to the record of decision was sought and approved. In the modification, a pump-and-treat system will be installed to treat only the most contaminated section of the plume, eliminating the problem of community disruption.

TABLE 4-2 Level of Participation of Stakeholders in Various Stages of the Superfund Process

Stage of Superfund Site Cleanup Process	Stakeholders						
	Public	Regulators	Potentially Responsible Parties/ Technology Users	Providers	Investors	Insurance Companies	Site Workers
Discovery	L	H	M	—	—	—	—
PA/SI	—	H	H	—	—	—	—
NPL Ranking	—	H	L	—	—	—	—
RI	L	H	H	—	—	—	—
FS	L	H	H	M-H	—	—	—
Public Comment	H	H	H	M	—	—	M
ROD	—	H	M	M	—	—	—
RD/RA	—	H	H	H	—	—	—
O&M	L	L-H	H	H	—	—	—

NOTE: "L" indicates low participation, "H" indicates high participation, and "M" indicates moderate participation. A dashed entry denotes no participation by the particular stakeholder group at the indicated stage of the cleanup process. See Figure 4-1 for definitions of the stages in the cleanup process.

BOX 4-4
New Bedford Harbor:
Citizen Opposition Halts EPA's Cleanup Plan

New Bedford Harbor, in southeastern Massachusetts, has been on the National Priorities List since 1982 because harbor sediments are contaminated with high levels of PCBs. Community concern about conventional remediation systems raises the possibility that innovative technologies may be used.

New Bedford, once a vibrant port for New England's whaling industry, now has a high unemployment rate and an eroded industrial base. The city has a large minority community, and many of the residents are of Portuguese descent, with limited English-speaking ability. These factors have raised questions about environmental equity in EPA's interactions with the community (Boston Globe, February 13, 1994).

The first phase of the harbor cleanup involves dredging a 5-acre hot spot and treating the contaminated soils. The average concentration of PCBs at the site is 30,000 parts per million (ppm); concentrations range from 4,000 to 200,000 ppm. At the high range, samples represent a virtually pure PCB product (P. Craffey, Massachusetts Department of Environmental Protection, personal communication, 1996).

EPA's 1992 proposal called for incinerating contaminated material excavated from the hot spot and placing incinerator residue in a capped landfill. Community opposition to the use of incineration was virulent and became the focus of intense exchanges between government and community activists. Opponents fear adverse health effects from incinerator emissions.

New Bedford City Council members accused EPA of bullying them into a cleanup plan they opposed. In September 1993, the city council passed ordinances prohibiting the transport of incineration equipment on

---

These concerns and expectations can be translated into success criteria that can be grouped into three categories: technical performance attributes, commercial attributes, and public and regulatory acceptance attributes, as described in the remainder of this chapter.

TECHNICAL PERFORMANCE

Technical performance attributes (see Table 4-1) comprise the first category of success criteria that can be used to evaluate the effectiveness of ground water and soil cleanup technologies. Technical performance attributes include the ability of the technology to reduce health and environmental risks by reaching desired cleanup end points. Also included in this category are a variety of factors

city streets and blocking electrical and water hookups to the incinerator site (New York Times, October 10, 1993). EPA in turn threatened the city with fines of $25,000 per day for impeding a federally mandated cleanup and secured a temporary order in the federal district court in Boston to block the ordinance preventing site access and utility service.

This downward spiral was halted when EPA, state and local officials, and community group representatives agreed to hire a mediator and to consider alternatives to the incinerator. The New Bedford Forum, a stakeholder group, was formed, and the group agreed on a set of criteria for selection of treatment technologies that included performance, availability of a unit to perform an on-site treatability study, cost, and past history.

The first step being taken under the revised plan is to conduct treatability studies. In early 1996, EPA announced that three vendors had been selected to perform treatability studies. The technologies to be tested are solvent extraction with chemical destruction; thermal desorption with chemical destruction; and in situ vitrification (P. Craffey, personal communication, 1996). These technologies were selected to represent a variety of possible approaches, and the EPA acknowledges that if none is sufficiently successful, another round of technology evaluation may be necessary. The dredged material is saturated with water and has approximately a foot of water on top, so all of the vendors may have to dry the material as a pretreatment step. The expectation is that each of the treatability studies will be performed for about a week at virtually full scale, thus minimizing the possibility of scale-up problems for the technology that is ultimately selected (P. Craffey, personal communication, 1996).

Although EPA makes the final decision, given the history of the project and the agency's current commitment to the New Bedford community, it is expected that the final technology selection will have wide support.

related to the ease of engineering the technology: robustness, forgiveness, ease of implementation, maintenance and down time requirements, predictability, ease of scaleup, and residuals production.

Health and Environmental Risk Reduction

The fundamental purpose of a remediation technology is to reduce risks to human health and the environment. However, the relative degree of risk reduction offered by one remediation technology versus another is very difficult to determine because quantitative estimates of health and environmental risks at contaminated sites are highly uncertain.

Major uncertainties exist in determining which populations have been exposed to contamination or are at risk of exposure. Table 4-3 and Figure 4-2 summarize potential exposure pathways. As shown in the table and the figure, there are multiple possible exposure routes. Predicting the degree to which each of these affects individuals near contaminated sites, or those farther from the sites who drink contaminated water or consume contaminated food, is a highly uncertain process, as illustrated by the complexity of the exposure pathways shown in Figure 4-2. Estimating contaminant concentrations to use as inputs to exposure calculations may involve significant uncertainty.

Adding to the difficulties in estimating the degree of health risk reduction achieved by a remediation technology is the fact that the complete profile of chemicals present at a waste site is frequently unknown. For example, after analyzing leachate at 13 representative sites across the country, the EPA was able to identify only 4 percent of the organic chemical constituents present in the leachates (National Research Council, 1991b). At many waste sites, a wide variety of compounds, such as solvents, fuels, and metals, may be present as mixtures. Determining the effectiveness of technologies against mixtures of chemicals, and the degree of health risk reduction that results when one element of the mixture is eliminated but another is not, can also be an uncertain process.

Further complicating estimates of health risk reduction, the toxicological properties of the contaminants, either singularly or when part of a mixture, may be uncertain or unknown. As of July 1993, the Registry of Toxic Effects of Chemical Substances contained 120,962 entries of chemicals known to have toxicological effects (Sweet, 1993). However, only 600 of the chemicals had undergone sufficient scientific evaluation to adequately document the specific effects on human health. Furthermore, health-based remediation decisions are based on cancer risks at most Superfund and other waste sites. However, populations are often exposed to contaminants at levels known to cause noncancer health effects such as low birth weight, birth defects, neurobehavioral problems, liver and kidney disease, cardiac anomalies, gastrointestinal distress, dermatological problems, headaches, and fatigue (ATSDR, 1994; National Research Council, 1991b). For example, the Agency for Toxic Substances and Disease Registry found that the 5,000 people listed in its registry for past exposure to trichloroethylene in drinking water reported higher than normal rates of diabetes, stroke, elevated blood pressure, and neurological problems (Johnson, 1993).

Other factors also complicate efforts to determine the health effects of exposure to contaminated soil or ground water and the benefits of reducing this exposure. Even when health studies are conducted, detecting health effects for which there is a long interval between contaminant exposure and the onset of disease may be difficult. Uncertainty also results from the possibility that the comparison groups (those not living near the waste site who are used to establish whether a higher than normal disease rate exists in the population under study) may also have been exposed to contaminants from some other source, such as the work place. Factors such as smoking, poor diet, and absence of prenatal and preventive

TABLE 4-3   Contaminant Exposure Pathways From a Toxic Waste Site

Medium	Exposure Pathway
Air	Inhalation of air vapor
	Inhalation of particulate matter
	Inhalation of soil vapor (for example, in basements)
Soil	Ingestion of soil
	Dermal contact with soil
	Ingestion of plants
	Ingestion of airborne soil
	Dermal contact with airborne soil
	Ingestion of airborne plant matter
	Ingestion of waterborne soil
	Dermal contact with waterborne soil
	Ingestion of waterborne plant matter
Ground water	Ingestion of ground water used as water supply
	Inhalation of vapors from ground water used as water supply
	Dermal contact with ground water used as water supply
	Inhalation of vapors from ground water in basements
	Dermal contact with ground water in basements
	Dermal contact with seepage water
	Inhalation of vapors from seepage water
	Ingestion of plants that take up contaminants from ground water
Surface water	Ingestion of surface water used as water supply
	Dermal contact with surface water used as water supply
	Inhalation of vapors from surface water used as water supply
	Inhalation of vapors from surface water used for recreation
	Dermal contact with surface water used for recreation
	Ingestion of plants irrigated with surface water
	Ingestion of aquatic biota

NOTE: Surface water may become contaminated via a variety of pathways, including runoff from the contaminated site, ground water from the contaminated site that discharges to the surface water body, and waste lagoon overflow.

SOURCE: Adapted from NYS DOH, 1993.

medical care may bias the results of health investigations (National Research Council, 1994). The net result of these uncertainties is that opinions about the risks posed by site contamination vary depending on who conducts the health investigation and who interprets the results.

Like determining human health risks, quantifying risks to the environment and the level of environmental risk reduction achieved by a given remediation process is very difficult, if not impossible. Effects of contaminants may vary

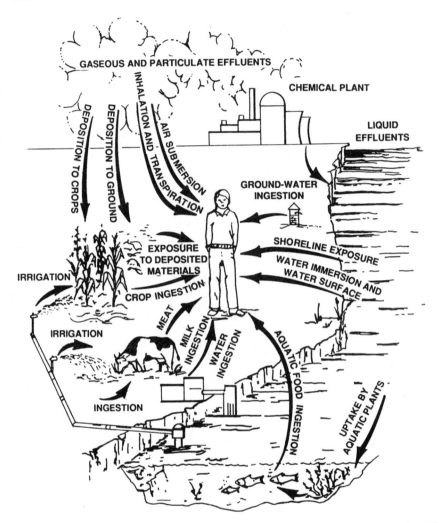

FIGURE 4-2   Pathways of human exposure to hazardous wastes. SOURCE: National Research Council, 1991a.

greatly among species. In addition, contaminants may bioaccumulate in the food chain: a small amount of contamination ingested by an organism low on the food chain can result in significant contamination in species at higher levels of the food chain. Further, wildlife risk assessments are based on the current limited knowledge of documented and/or predictable outcomes, such as tumor growth or subsistence capability. These assessments customarily consider only the ability of a specific species to survive, as opposed to evaluating the health of the species.

Using this approach may result in failure to observe a host of negative outcomes, such as reduced ability to reproduce, DNA anomalies, multigenerational effects, teratogenic effects, increased contaminant body burdens, and so on.

The public reaction to all of this uncertainty in risk assessment is complex. People who believe they have experienced health effects from exposure to contaminated sites may feel stigmatized or discredited, at both the individual and community levels (Edelstein, 1992). In addition, citizens often feel alienated and betrayed by government agencies when their concerns cannot be addressed and questions cannot be answered definitively. These public reactions can lead to distrust of agencies and rejection of remediation technologies for which exact performance and ability to meet applicable regulatory standards cannot be guaranteed prior to use, especially when local residents have not been involved early in the site investigation and remediation process.

Given the unknowns in fully defining the human health and environmental effects of contaminants in ground water and soil, the dilemma is how to define remediation technology performance in a way that is both quantifiable and relevant to the goal of preventing adverse effects. Obviously, under the ideal scenario the technology would eliminate exposure to the contamination by removing all of it from the site, or, for contaminated ground water, it would remove enough contamination so that the water meets regulatory standards for drinking water. As explained in Chapters 1 and 3, achieving these standards may not be feasible at complex hazardous waste sites. In addition, regulatory standards for ground water and soil remediation vary depending on the state in which the site is located, the regulatory program under which the site is governed, and the individual regulator who is in charge of overseeing the site. These technical limitations and regulatory variations make it very difficult to decide which standards should be used to evaluate the performance of a technology.

Human health and environmental effects of contaminants are the result of exposures of populations to a contaminant of sometimes unknown toxic effect through a pathway. To be successful, a technology must be capable of reducing the toxicity and/or intercepting or eliminating the movement of the contaminant along the pathway in some manner so that receptors are unharmed. Therefore, regardless of the regulatory program under which a site is administered and regardless of the degree of complexity of the contaminated site, the following four criteria should be used in describing a technology's ability to reduce risks posed by the contamination:

1. *Reduction in contaminant mass.* The quantity of contaminants in the subsurface provides the best indication of the potential longevity of the ground water and soil contamination problem, and removing contaminant mass from the subsurface reduces long-term exposure to the contaminant due to transport through subsurface pathways.

2. *Reduction in contaminant concentration.* Reducing contaminant concen-

trations reduces the level of exposure to contaminants and thereby diminishes the risk associated with that exposure. Reductions in contaminant concentration often parallel reductions in contaminant mass, and contaminant mass is usually estimated form concentration measurements. However, it is important to recognize that contaminant concentrations may decrease (due to mass transfer and dilution process) while the total mass of contaminants remains unchanged.

3. *Reduction in contaminant mobility.* Remediation technologies that immobilize the contaminants can effectively eliminate exposure risk even if the contaminants remain in the ground—as long as people are kept off site and are prevented from using contaminated water. The fundamental performance measure for remediation technologies that reduce contaminant mobility is their ability to prevent contaminants from returning to the zones of natural ground water flow.

4. *Reduction in contaminant toxicity.* Remediation technologies can reduce contaminant toxicity by converting the contaminants to a less toxic or less bioavailable form. The total concentration of contaminants in a soil or waste deposit may have little to do with overall toxicity (Chaney and Ryan, 1994; Alexander, 1995; Ruby et al., 1996), because the toxicity of contaminated soil depends upon the bioavailability of the material when ingested. Surrogate methods such as leaching tests are commonly used to estimate the bioavailability of soluble toxic compounds in soil.

When using these four measures to assess risk reduction, a key question to consider is at what point the reduction in contaminant mass, concentration, mobility, or toxicity should be measured. For ex situ soil or ground water processing technologies, the answer is self evident: at the exit from the processing unit. However, for in situ ground water remediation technologies, the answer often depends upon the site situation and the technology. Often, technologies do not act immediately and completely at the point of application. For example, pump-and-treat systems may withdraw contaminants at the pumped well but also reverse or halt the migration of contaminants at a distance from the well. Biotreatment systems reduce contaminant concentrations along the contaminant flow path, downgradient from the point of introduction of substances that encourage growth of contaminant-degrading organisms. On the other hand, a permeable reaction wall may be installed close to the source of contamination or at a prescribed distance from the contaminant source depending upon the source strength, plume configuration, and cost of installation. In many cases, the location at which the technology is installed will be dictated by the capability of the technology and an assessment of the most cost-effective configuration. The configuration is also influenced by the environmental goals to be met at a point of compliance defined by a regulatory agency. The technology user and the regulator must agree on the appropriate location for monitoring technology performance: within the contaminated area, at its boundary, or somewhere in between, depending upon the technology and site-specific situation. This variation in the location at which technology performance is measured can make it difficult to standardize performance reporting and to compare technologies. When comparing technologies, it may be necessary to specify the performance at the point of maximum effect and the distance of that point under some known (or standardized) flow or residence time condition.

All of these considerations must enter into formulating data needs and testing protocols used to judge the effectiveness of a technology.

Engineering Friendliness

Remediation technology performance is often described solely in terms of the ability to reduce contaminant concentration or mass in the subsurface. However, the ease of designing and operating the technology—its "engineering friendliness"—can be the critical factor in determining whether a technology becomes widely used. Key factors related to engineering and operation of technology are robustness, forgiveness, ease of implementation, maintenance requirements, predictability, and residuals production.

Robustness

Robust remediation technologies are effective over a range of contaminant and site conditions. The operating environment, including not only the geologic factors described in Chapter 1 but also ambient temperature, volume and acidity of rainfall, soil moisture, and pH, varies widely across the country and with the season. Environmental factors may have a significant effect on the cost and performance of a remediation technology. Furthermore, experience has shown that site characterization information can be inadequate in describing site parameters and contaminant characteristics (National Research Council, 1994) and that technologies that perform optimally under a narrow range of conditions may not perform well under the particular conditions actually encountered once remediation begins. Waste stream variability is a challenge for off-site hazardous waste treatment facilities: despite the implementation of waste analysis plans and acceptance of discrete batches from known generators, receipt of incompatible waste has caused serious accidents, for example at solvent recovery operations. When cleaning up waste in the ground, challenges associated with waste stream variability are greatly magnified. A robust technology can operate effectively in a wide range of ambient operating conditions and can tolerate a wide range of variability in waste characteristics.

Forgiveness

Forgiveness describes the extent to which the remediation technology is sensitive to operating conditions. A forgiving technology will meet remediation goals even if there are fluctuations in the way the process is operated. If, for example, operating temperatures or reactant feed rates or residence times do not have to be precisely controlled, the technology will have a greater chance of success under field conditions. Many of the successful soil and ground water cleanup technologies in use today (including vacuum extraction, air sparging, and aerobic bioremediation) are forgiving in nature. If a technology operates effectively only within a narrow range of conditons, its widespread use may be limited.

Ease of Implementation

Remediation technologies have variable requirements for site preparation and technology construction. Infrastructure needs, such as access to highways, electricity (especially if large amounts are needed), clean water, and wastewater treatment facilities, will vary with the technology and also with the site. Required operator skill levels and number of site workers also vary greatly among different remediation technologies. For example, for a process requiring rapid analysis of site data, using a field-portable gas chromatograph/mass spectrometer (GC/MS) may allow use of a dynamic work plan and thus yield time advantages over conventional remediation technologies. One of the tradeoffs, however, is that the field-portable GC/MS requires a highly skilled operator. When technologies require very specialized skills, delays in implementation can result from difficulty in obtaining an expert's time. The amount of time required and skills needed to prepare a site, move equipment to the location, make it operational, and then dismantle it when remediation is complete are criteria that decisionmakers take into account.

Duration of treatment is also a factor in establishing ease of implementation. Remediation technologies (such as conventional pump-and-treat systems) that require long periods of operation and maintenance present accountability and logistical challenges that are vastly different from those of a technology that requires little or no long-term care.

Maintenance and Down Time

Remediation technologies vary in their requirements for maintenance during remedial operations. For some technologies, scheduling down time in advance may be possible, creating an understanding that preventive maintenance is needed. Technologies that break down during operation not only cost valuable time but also may create an image of unreliability to clients, regulatory agencies, and the interested public, who may be scrutinizing remediation operations. A reliable technology is one that performs without down time during remedial operations or one for which preventive maintenance can be anticipated and scheduled at the beginning of operations.

Predictability and Ease of Scaleup

For innovative technologies that have undergone limited field applications, questions associated with predictability under the wide range of site conditions that might be experienced have an impact on commercial viability. Parameters identified above under robustness and forgiveness may affect the ability of a technology to perform effectively under different conditions.

Ease of scaleup refers to the ability of a technology that has performed well under bench and pilot conditions to be designed to perform at full scale. Some technologies that are promising in the laboratory or are effective in small experimental applications may not operate effectively at full scale in the field. This failure during scaleup may occur because the basic understanding of the critical processes governing technology performance is lacking or because laboratory hardware cannot be translated into field-size equipment. For in situ processes, large-scale variations in geologic structure may not be accounted for in smaller-scale experiments, causing unexpected problems in full-scale implementation. Uncertainty with respect to scaleup affects the viability of a technology for commercialization.

Secondary Emissions and Residuals Production

Many treatment technologies can produce secondary waste streams with their own risks. Examples are sludges from metals precipitation or volatile organic compounds from air stripping or soil vapor extraction. The quantity and character of these waste streams may greatly affect the environmental acceptability of the technology and the cost of using it. Often the technology user will carefully review a technology to determine if there will be unexpected costs associated with the disposal of secondary wastes. Regulators also review the potential for environmental problems that may arise from secondary waste generation. Minimization of secondary emissions is often a strong driver in the community affected by the remediation effort. The potential for producing health hazards, odors, or dust and the need for disposal of additional wastes at the remediation site can lead to strong community opposition. Consequently, the extent and character of secondary wastes from a remediation process must be understood and weighed against stakeholder concerns.

COMMERCIAL ATTRIBUTES

The second general category of criteria important in evaluating ground water and soil cleanup technologies is a technology's commercial attributes. Cost and potential for profit are critical considerations in the commercialization of any product, and waste-site remediation technologies are no exception. The commercial value of a technology depends on its cost competitiveness (its capital and operating costs relative to those of other technologies used for the same purpose), the intellectual property rights that the supplier can hold secure, and, most importantly, the potential profits the technology may generate.

Capital and Operating Costs

The remediation technology provider must know the cost of the product in order to determine the market potential and the margin for profit. Cost is one of the most important factors in technology selection (environmental protection between alternatives being equal) for the private user and is also a factor that, by law, regulators must consider. According to the Superfund act, the primary criteria that must be balanced in remedy selection are long-term effectiveness and permanence; reduction of toxicity, mobility, or volume through treatment; short-term effectiveness; implementability; and cost (U.S. Government, 1990a). Similar factors must be considered under RCRA (U.S. Government, 1990b).

Stakeholders in the use of a remediation technology need to have a clear understanding of the full costs (capital and operating) of implementing that technology. The user and the regulator must know the cost in order to choose the most cost-effective remedy. The user must also know the cost to obtain the authorization necessary for funding. The technology provider needs cost data in order to determine if the technology will be competitive in the marketplace and ultimately to be able to sell it to a potential user. Uncertainty about the competitive niche of a technology can be a barrier to development or can lead to large expenditures for development of a technology that cannot compete. Consequently, having a good indication of the cost of a technology is important very early in the development cycle.

When a technology becomes available and ready for use, decisions are made about the cost effectiveness of the technology in a particular application. In order for a new technology to be considered for implementation at a site, consistent and reliable cost information must be available to decisionmakers. Traditionally, this information has been supplied by the technology provider or the consultant designing the remedy. Information in technical literature or from the technology provider is reported in many formats, which often makes it difficult to compare costs of competitive technologies without substantial development of information by the consultant to allow side-by-side comparisons. In addition, there is often a strong site-specific element to the technology cost, but often costs are presented with insufficient information to transfer technology and cost information between sites.

Uniform cost reporting is an essential element to facilitate the comparison of technologies and to speed the acceptance of new competitive technologies. The EPA has recognized the importance of cost in developing its Superfund Innovative Technology Evaluation Program (see Chapter 5) for testing innovative technologies. The "applications analysis reports" developed under this program have attempted to standardize cost reporting formats to allow for critical analysis of costs. Similarly, other federal agencies, which are both users and developers of remediation technology, have recognized the need for consistency in cost reporting and are developing uniform cost and performance reporting guidelines under the auspices of the Federal Remediation Technologies Roundtable (Federal Remediation Technologies Roundtable, 1995). The Department of Defense is also trying to standardize cost reporting to aid in screening technologies. Despite these efforts, greater standardization is needed in the reporting of capital and operating costs, as is discussed in detail in Chapter 6.

Intellectual Property Restrictions

The impact of copyright or patent restrictions on a remediation technology's market potential varies depending on the interests of various stakeholders and their involvement in the decisionmaking process. Nevertheless, it is possible to make a few generalizations. The technology provider wants to have exclusive rights to market or license the technology for as long as possible in order to maximize the profit on the investment made in developing the technology. A patent provides its owner with the right to exclude others from practicing the invention for up to 20 years, thus allowing the inventor or developer to obtain a profit. An investor looks for a strong position in intellectual property rights when deciding whether to fund technology development. Technology users, regulators, and the public want enough information about the technology to be assured that it will perform effectively and efficiently. A patent sometimes acts to provide some of that security, but it may also limit access to the technology for several years. In addition, to the extent that proprietary interests may cause a technology supplier to withhold some information, the public may have less trust in a technology that is based on trade secrets than in one for which all available information is freely shared. Because trust is an important factor in risk communication, proprietary interests may affect the viability of a technology's public acceptance.

Profitability

Expectation of profit drives research and development as well as investment in innovative technologies. Continuing realized profit is the engine that maintains the ongoing use of a remediation technology and funds continued research and development. As discussed in Chapter 2, an investor will seek opportunities to fund technologies that provide the greatest potential financial return. Stakeholders other than the investor and technology provider have less interest in the profitability of a technology, except that they may want evidence that the technology vendor is sufficiently solvent to complete the remediation on a reasonable budget.

Accessibility

For a variety of reasons, a remediation technology may be inaccessible to potential users. In some cases, patents restrict access to the technology. For example, universities are increasingly holding the intellectual property rights to technologies they have developed in the hopes of supporting their programs through profits and licensing, but often they lack the administrative, legal, and financial structures to follow through with the details needed to support the license. In other cases, technologies are developed and used within the government sector, but because of poor communication or inability to transfer the technology to other users, the technology may not be easily commercialized. In some cases, small private firms may offer technologies that they have used successfully in a small geographic area but do not have the funding, staffing, or desire to increase the technology's use. If access to a technology is restricted, choosing the innovative approach over other, more accessible methods may be too time consuming and expensive for a potential technology user with an immediate remediation need.

PUBLIC AND REGULATORY ACCEPTANCE ATTRIBUTES

The third major set of technology performance measures includes attributes that, while not always quantifiable, may greatly influence whether a remediation technology is selected for use in a specific situation. Public and regulatory issues are intertwined in that public concerns often translate into regulatory action to meet the concerns. The types of issues that are most important to the public vary with the waste site. Major issues of concern to the public and regulators, in addition to those already discussed in this chapter, may include disruption to the community or ongoing activities at or near the waste site, safety of the remedy, whether using the technology will slow the cleanup by creating new regulatory hurdles, and usability of the land once cleanup is complete.

Disruption to the Community

Disruption to the community in implementing a technology may be a significant barrier to the acceptance of a technology in specific instances. Such disruption varies depending on the technology and can take the form of traffic increases, visual impacts, noise, odors, and load on the local wastewater treatment system. Other aspects of disruption relate to loss of natural resources or water supplies that may occur even though remediation is undertaken (see Box 4-2).

Another form of disruption relates to changes in the social fabric of the community that may arise from identification of a contaminated site, dissention over selection of remediation technology, and impacts on perceptions of the community and property values. On occasion, questions may arise about the need to employ a cleanup technology in light of the community disruption that may occur, and the remedy may be modified to balance environmental needs against community concerns (see Boxes 4-3 and 4-4).

When remediation technology choice becomes the subject of debate, impacts will be evident in the editorial pages of newspapers and the outcomes of local elections and may affect long-standing relationships among neighbors.

Disruption to Ongoing Site Activity

Contaminated sites vary widely in their proximity to other ongoing activities, either on the site or in close proximity. Commercial operations under the same or different ownership may be taking place on the same site. Where there are ongoing operations, the choice of a particular remediation technology may be optimized for short duration of cleanup time or minimal visual impact, excavation activity, or noise. Technologies that involve excavation and transportation of large volumes of contaminated material will require attention to traffic patterns and timing to minimize conflict between the remediation activity and ongoing site operations. If the site is very small, a technology requiring few staff and modest equipment may be favored in order to avoid disrupting commercial or other activities. Activities and organizations off the site but in close proximity, particularly those such as schools or hospitals that involve sensitive receptors, also may be important factors in technology selection.

Safety

The perceived safety of a remediation technology to nearby inhabitants and the safety of workers performing the remediation can both affect decisions about whether to use the technology.

A community's perception of the safety of implementing a remediation technology can be crucial to the use of a technology. The concern about possible health risks from remediation activity often relates directly to secondary emissions or residuals. The community may seek assurances that any cleanup actions will not result in transfer of risks or creation of new risks. For example, citizen groups often are concerned that remediation involving incineration may result in generation of new pollutants as products of incomplete combustion (see Box 4-4).

Workers at remediation sites are under the jurisdiction of the Occupational Safety and Health Act and are required to undergo training in personal protection and emergency procedures. Some technologies are inherently more dangerous to workers than others. Implementation of technologies requiring excavation with heavy equipment, use of high voltage electricity in damp areas, and demanding physical labor by individuals wearing full protective equipment offer a few examples of situations that may result in injuries. In situations where remediation is occurring on a site along with other manufacturing or commercial activities, safety considerations may affect a larger number of people.

To the extent that a user does not want to be associated with an unsafe technology and the provider does not want the liability problems associated with safety incidents, innovative technologies of questionable safety will not be favored.

Regulatory Hurdles

Many of the regulatory hurdles in technology acceptance are related to the attributes described above, because the regulator will endorse the use of a technology after reviewing the technology for its merits and deficiencies. However, other purely regulatory issues can also influence technology acceptance. These issues are primarily related to whether or not in applying the technology, the many operational and performance constraints imposed by regulation can be met without excessive complication or cost.

The Superfund regulatory process and its state equivalents are targets for criticism. Members of the public believe that their comments on proposed cleanup actions and technology choices are solicited too late in the process. Technology developers, owners, and users believe that the process takes too long and the outcomes are too unpredictable to drive markets for technology or to allow technology consumers to apply sound business principles in making technology choices. Providers of technology services are cognizant of regulatory hurdles for innovative technology, and this may explain why consultants often recommend that their clients use proven technology and avoid the potential for delay and the uncertain risks that may be associated with innovative technology.

Regulators have recognized that regulatory barriers may be playing a role in constraining the use of innovative technology and are developing a variety of programs to address perceived problems associated with innovative technology testing and demonstration (see Chapter 5).

Future Land Use

Use of land following remediation is a particularly important concern for the public. It is difficult for the general public to understand why the government spends millions of dollars to clean a site that still must be fenced in to prevent access. Technologies that produce residuals requiring strict limits on future land use will be less attractive to the public and site owners than technologies that require few limits.

In brownfield areas, sites where current use is unattractive or prohibited because of contamination are revitalized through remediation to create new possibilities for land use, including commercial and manufacturing activity, without necessarily restoring the site to its preindustrial condition (see Chapter 1). In brownfield areas, site owners and future land users have an interest in technologies that achieve cleanup goals quickly and require few restrictions on the future ability to use the land for industrial or commercial purposes.

CONCLUSIONS

The overlapping interests and divergent expectations of stakeholders in contaminated site remediation form a complex web of shared concerns. In general, the affected public wants to "fix the problem irrespective of cost," whereas technology users wish to "fix the problem at the lowest possible cost." This divergence of views may explain a great deal of the acrimony that is associated with site cleanup decisions. Nevertheless, these two stakeholder groups have substantial common interest in remediation technology performance criteria such as ability to achieve risk reduction, predictability of the technology, safety, and disruption to the community and the site. Regulators act as a bridge between the affected public and technology users. Investors, insurance companies, and site workers also have interests that must be addressed to ensure successful technology application.

Achieving success in remediation depends on effectively addressing various stakeholder expectations in order to emphasize the points of common agreement. Any number of stakeholders can prevent the use of a remediation technology, so there must be substantial common ground in acceptance for a technology to be successful. Key factors that technology developers must consider when testing a new technology and evaluating its market potential can be organized in three categories: technical attributes, commercial attributes, and public and regulatory acceptance attributes. It is imperative that reliable and consistent information be made available in all these areas to all parties to allow the stakeholders to evaluate the acceptability of innovative remediation technologies.

RECOMMENDATIONS

To streamline the process of remediation technology selection and minimize acrimony among stakeholders at contaminated sites, the committee recommends the following:

• **The EPA and state environmental regulators should amend their public participation programs and require that public involvement in contaminated site cleanup begin at the point of site discovery and investigation.** An informed public is better prepared to participate in the review of technology selection options and to consider innovative remediation technologies. Once site data are collected, the data should be made available at a convenient, accessible location of the public's choosing. While some members of the public desire short, factual data summaries, others may have expertise that equips them to review and evaluate the full studies, including laboratory analytical data and study protocol. To further assist the community, sources of toxicological and health information on contaminants of concern, as well as technical data collected from other sites where different technologies have been implemented and assessed, should also be provided. Anecdotal evidence suggests that innovative remediation technologies are selected more frequently when the public is involved early in the site remediation process.

- **The EPA should work to eliminate the preference for a linear process in remediation technology selection.** Under Superfund, amending the ROD to change the remediation technology to reflect new data and advances in technologies is a cumbersome process. The process should be streamlined to allow for application of innovative remediation technologies if new performance data indicate that an innovative remedy is a better choice than the original remedy.

- **Technology developers should report the effectiveness of their systems in reducing public health and environmental risks based on the technology's ability to reduce contaminant mass, concentration, mobility, and toxicity.** At complex sites, no technology can entirely eliminate all risks associated with contamination. In addition, determining the link between environmental contaminants and health and environmental effects is a highly uncertain process because of unknowns related to contaminant toxicity and exposure pathways. A technology can only reduce risk by reducing the magnitude and duration of the exposure of a target receptor to a contaminant. Consequently, these measurable, technology-specific criteria must be used as surrogates for environmental and health effects, regardless of the regulatory program under which the contaminated site is administered. Technology developers should report the range of uncertainty in these measured values to allow for meaningful comparisons of risk reduction potential offered by different technologies.

- **Technology developers and suppliers should specify the performance of a remediation technology at the point of maximum effect and should specify the distance of that point from the application of the technology under some known or standardized flow or residence time condition.** Depending on the technology and how it acts in the field, the full effect of a technology may occur at some distance from the actual point of application. Specifying the point of maximum effect and its distance from the technology installation will improve comparison of remediation technologies.

- **Technology developers should consider public and regulatory concerns about remediation technology use when testing remediation technologies.** While not subject to quantitative measures, public and regulatory concerns are important to technology acceptance.. Even if a technology meets technical and commercial measures of success, strong public or regulatory objections may make it undesirable. Often these concerns center around site-specific debates, such as disruption to the community, but they may surface often enough that a particular technology is at a disadvantage.

REFERENCES

Alexander, M. 1995. Critical review: How toxic are toxic chemicals in the soil? Environmental Science & Technology 29(11):2713-2717.

ATSDR (Agency for Toxic Substances and Disease Registry). 1994. Biennial Report to Congress, 1991 and 1992. Atlanta: U.S. Department of Health and Human Services, ATSDR.

Boston Globe. 1994. The EPA's dirty secret. February 13, Editorial Section, p. 74.

Brown, P., and E. Mikkelson. 1990. No Safe Place: Toxic Waste, Leukemia, and Community Action. Berkeley, Calif.: University of California Press.

Chaney, R. L., and J. A. Ryan. 1994. State-of-the-art in evaluating the risks of As, Cd, and Pb in urban soils for plants, animals and humans. Paper presented at the Conference for Decision Finding in Soil Protection: Evaluation of Arsenic, Lead, and Cadmium in Contaminated Urban Soils, Braunschweig, Federal Republic of Germany, October 9-11, 1991.

Edelstein, M. 1992. Psychological impacts in contaminated communities. Paper presented at Agency for Toxic Substances and Disease Registry and New York State Department of Health Workshop, Rensselaerville, N.Y., June 8-9, 1992.

EPA. 1996. Innovative Treatment Technologies: Annual Status Report (Eighth Edition). EPA 542-R-96-010. Washington, D.C.: EPA, Office of Solid Waste and Emergency Response.

EPA Region I. 1995. Administrative Order by Consent for Additional Remediation Investigation/ Additional Feasibility Study, Appendix A, Statement of Work, Additional Remedial Investigation and Feasibility Study, Phase II. EPA-CERCLA Docket No. I-95-1048. Boston, Mass.: EPA Region I.

EPA Region II. 1993. Explanation of Significant Differences, Caldwell Trucking Company Site, Fairfield Township, New Jersey. New York: EPA Region II.

Ewusi-Wilson, I., P. Foti, C. Jepsen, S. O'Brien, and D. Pena. 1995. Technology Evaluation for Remediation of PCB-Contaminated Sediment in the Housatonic River. Medford, Mass.: Tufts University, Department of Civil and Environmental Engineering.

Federal Remediation Technologies Roundtable. 1995. Guide to Documenting Cost and Performance for Remediation Projects. EPA-542-B-95-002. Washington, D.C.: EPA.

Gallagher, D. 1993. Town fights proposal for Superfund site. Los Angeles Times (May 22), sec. B1, p. 5.

GEI Consultants, Inc. 1995. Additional Feasibility Study: Initial Screening of Remedial Alternatives, Pine Street Canal Site. Burlington, Vt.: GEI Consultants, Inc.

Johnson, B. L. 1993. Testimony before the Subcommittee on Superfund, Recycling and Solid Waste Management, U.S. Senate, Washington, D.C., May 6, 1993.

Kraus, N., T. Malmfors, and P. Slovic. 1992. Intuitive toxicology: expert and lay judgments of chemical risks. Risk Analysis 12(2): 215-232.

McCoy and Associates. 1993. Commercialization of innovative treatment technologies: a ten-year perspective. The Hazardous Waste Consultant (May/June):4.1-4.22.

Miller, C. 1995. Superfund coalition against mismanagement. Testimony before the Subcommittee on Water Resources and Environment Committee on Transportation and Infrastructure, U.S.House of Representatives, Washington D.C., June 21, 1995.

National Research Council. 1991a. Environmental Exposure: Report of a Symposium. Washington, D.C.: National Academy Press.

National Research Council. 1991b. Public Health and Hazardous Wastes, Environmental Epidemiology, Vol. 1. Washington, D.C.: National Academy Press.

National Research Council. 1994. Alternatives for Ground Water Cleanup. Washington, D.C.: National Academy Press.

New York Times. October 10, 1993. Pollution remedy is hotly debated. A28.

NYS DOH (New York State Department of Health). 1993. Exposure Pathways From a Toxic Waste Site. Albany, N.Y.: NYS DOH.

O'Donnell, M. J. 1995. Pine Street Canal Superfund site, Burlington, Vermont: Disapproval with modifications required of the additional feasibility study initial screening of remedial alternatives report, September 8, 1995. December 4. EPA Region I, Boston, Mass. Letter to M. L. Johnson.

Ruby, M. V., A. Davis, R. Schoof, S. Eberle, and C. M. Sellstone. 1996. Estimation of lead and arsenic bioavailability using a physiologically based extraction test. Environmental Science & Technology 30(2):422-430.

Stuebner, S. 1993. Triumph's Test of Will. Idaho Falls Post Register (July 25), news section.

Sweet, D. V. 1993. Registry of Chemical Substances Comprehensive Guide to the RTECS. Atlanta, Ga.: U.S. Department of Health and Human Services, ATSDR.

U.S. Government. 1990a. Federal Regulation 145, 40 CFR 300.430(f)(1)(i)(B), July 27.

U.S. Government. 1990b. Federal Regulation 145, 40 CFR 264.525(b) - proposed 55, July 27.

5

Testing Remediation Technologies

Many major U.S. industries, such as the pharmaceutical and automotive industries, have standard protocols for testing new product performance. Such protocols are lacking in the hazardous waste site remediation industry. This lack of protocols contributes to the difficulties that remediation technology developers face in trying to convince potential clients that an innovative technology will work. Lacking performance data collected according to a standard protocol, clients may hesitate to choose an innovative remediation technology because of the uncertainty in how the innovative technology will perform in comparison to a conventional technology. The types of data collected for evaluating remediation technology performance vary widely and are typically determined by the preferences of the consultant responsible for selecting the technology, the client, and the regulators overseeing remediation at the contaminated site. From the perspective of the client and the service providers who are interested in solving the immediate problem in a cost-effective manner, such a site-specific strategy is justified. However, from the broader perspective required for remediation technology development and testing, the performance and cost data needed to meet site-specific objectives are often insufficient to extrapolate the results from one site to another.

As a result of the lack of standard procedures for remediation process testing, many of the early attempts at soil and ground water cleanup, especially at complex sites, served as poorly planned and very costly national experiments. Expensive remediation systems were installed to clean up sites with very little understanding of the mechanisms controlling their performance. The results of these efforts were evaluated to try to gain a better understanding of mechanisms governing remediation, but such evaluations were complicated by the lack of standardized data sets (National Research Council, 1994; EPA, 1992). In some cases, remediation systems, such as soil vapor extraction (SVE), proved successful despite the limited understanding. In other cases, however, such as with pump-and-treat systems, tens of millions of dollars were spent at individual sites to install systems that later proved unable to meet cleanup goals (National Research Council, 1994; EPA, 1992).

Since the early 1990s, the Environmental Protection Agency (EPA) and other federal agencies have increasingly recognized the limitations of existing data on remediation systems and have taken steps to improve the consistency of data collection at contaminated federal sites. In 1995, the Federal Remediation Technologies Roundtable, a group of lead agency representatives involved in site remediation, issued guidelines for the collection of remediation cost and performance data at federal facilities (Federal Remediation Technologies Roundtable, 1995). Nevertheless, no standard process exists for data collection and reporting at privately owned contaminated sites, and the degree to which the Federal Remediation Technologies Roundtable guidelines are applied at federal facilities is unclear. The challenge for remediation technology development is to provide a framework and an infrastructure so that the individual benefits accruing to service providers and clients at specific sites, both federal and private, are gradually aggregated. Aggregation and critical review of data gathered according to standard protocols at numerous sites are essential for ensuring that the data are widely accessible to other technology developers and users, so that the success stories are not derived solely from anecdotes or unpublished reports.

This chapter describes a set of general principles that should be applied when testing performance of remediation technologies. It outlines the types of data needed to prove the performance of different classes of technologies, how to choose an appropriate test site for a remediation technology, and how to determine the amount of additional testing required to evaluate whether a technology tested at one site is applicable at another site. It also recommends ways that policymakers and others can encourage standardization in the collection of data on remediation technology performance.

DATA FOR PROVING TECHNOLOGY PERFORMANCE

Commercialization is the process of increasing use of a technology to solve a particular problem. Those who are considering use of an innovative remediation technology early in the commercialization process must decide whether the benefit (performance) of the technology is commensurate with its risk (failure to attain regulatory requirements). Generally, the user's greatest concern is having to do more: apply the technology over a longer period, implement an additional technology, or abandon the innovative technology and apply a conventional one. Therefore, to commercialize a remediation technology, the technology developer must convince prospective users that the innovative technology will cost effectively solve their problems with minimal risk (cost) of failure.

There are two approaches to minimizing the risk of using an innovative technology. The first is to guarantee performance. Such a guarantee requires assumption of financial risks or of residual liability. If the technology fails, the seller of the technology assumes the cost of meeting the remedial goals or the liability for noncompliance. To be able to offer the guarantee, the seller must have sufficient assets to make the guarantee credible. Given their limited financial resources, this is not possible for many technology developers. The second approach to minimizing the risk of using an innovative technology is to provide enough data so that the user is confident in the ability of the technology to provide the desired result. The data must be sufficient to verify the technology—that is, to prove its performance under a specific set of conditions with assurance of data quality.

The data required to verify performance include proof that the technology works under field conditions and proof that the technology will be accepted by regulators. In order to prove that the technology works to the satisfaction of potential clients and regulators, the technology developer will need evidence to answer two fundamental questions:

1. Does the technology reduce risks posed by ground water or soil contamination? That is, what are the levels of risk reduction achieved by implementing the technology?

2. How does the technology work in reducing these risks? That is, what is the evidence proving that the technology was the cause of the observed risk reduction?

As described in Chapter 4, remediation technologies reduce risk by decreasing the mass, concentration, mobility, and/or toxicity of contaminants in the subsurface. Direct measurements showing decreases in one or more of these parameters are essential for proving technology performance, but they are not sufficient to prove that the technology was responsible for the observed decrease in contamination. For example, contaminant concentrations in ground water may decrease for a variety of reasons, including sorption of contaminants by soil or aquifer solids, dilution due to natural mixing with uncontaminated ground water, biodegradation by native soil microbes, or chemical reactions with substances naturally present in the subsurface. A cause-and-effect relationship between application of the remediation technology and observed decreases in contamination must be established by collecting data to answer the second question, how does the technology work? Without answering this question and understanding the mechanisms responsible for performance of the technology, the technology design cannot be optimized, and the technology cannot be reliably transferred to other sites. In the past, technology tests have rarely been performed using protocols that answer this second question. This failure to gather evidence to explicitly link performance to remedial process has slowed regulatory acceptance and site-to-site transfer of innovative remediation technologies.

Demonstrating Risk Reduction Achieved by the Technology

To answer the question of whether the remediation technology works in reducing health and environmental risks, field tests are required to determine the reductions in contaminant mass, concentration, toxicity, and/or mobility achieved after application of the technology. Demonstrating reductions in all four risk measures—mass, concentration, toxicity, and mobility—is not necessary. Rather, the technology evaluation should provide two or more types of data leading to the conclusion that the technology has succeeded in decreasing one or more of the four risk measures. Which measure is appropriate depends on the remediation end points that the technology is designed to achieve.

Contaminant concentrations in the field following application of a remediation technology are readily determined by analyzing ground water samples from monitoring wells and soil samples from soil cores according to standard procedures. Likewise, decreases in contaminant mobility can be documented through standard tests that analyze contaminant leachability (although these tests are sometimes misapplied). However, documenting reductions in contaminant mass and toxicity is more challenging.

Quantifying contaminant mass in the subsurface, both before and after remediation, can present a significant challenge due to the complex distribution of contaminants among different phases (dissolved, sorbed, nonaqueous liquid, or solid) in both the horizontal and vertical directions. Contaminant mass is typically estimated based on concentration data from monitoring wells and soil core samples and on an estimation of the volume of contaminated material (mass equals concentration multiplied by volume). For example, in a field experiment to evaluate intrinsic remediation of petroleum hydrocarbons, the mass of hydrocarbons remaining in the subsurface at any given time was estimated by integrating concentration data from a network of monitoring wells over the contaminated area (Barker et al., 1987). However, although contaminant concentration and contaminant mass are closely linked and although contaminant mass is usually estimated based on measures of concentration, a reduction in contaminant concentration does not always signal a reduction in contaminant mass. Contaminant concentrations may decrease due to a manifestation of rate-limiting mass transfer phenomena or due to dilution with uncontaminated waters, while the total mass of contaminants remains essentially the same. The uncertainties associated with estimating total contaminant mass based on concentration data from discrete sampling locations at a heterogeneous site are often not reported.

Determining the toxicity of contaminants in the field is likewise difficult because of the cost and complexity of the studies required to link contaminant exposure to human health and ecological damage. The actual toxicity of contaminants to both human health and ecosystems can be measured only through long-term studies that assess the health and ecological impacts of contaminants. Such studies exist for some contaminants but not for others (see Chapter 4). An alternative for contaminated material that has been solidified or stabilized is to use leaching tests that analyze for toxic compounds in water that might leach through the solidified or stabilized material. Test methods for assessing the toxicity of leaching water include the extraction procedure toxicity test and the toxicity characteristic leaching procedure (EPA, 1989).

Methods for measuring decreases in contaminant mass and concentration differ somewhat depending on whether the remediation technology is designed to stabilize or contain contaminants, or to extract or destroy them. For stabilization and containment technologies, decreases in mobile contaminant mass should be determined by analyzing the amount of contamination available for transport to zones of natural ground water flow; for all other types of technologies, decreases in mass should be determined by analyzing the amount of mass remaining within the zone of remediation. For stabilization and containment technologies, effects on contaminant concentration should be determined by analyzing concentrations outside the zone of remediation, while for other types of technologies concentration or mass decreases should be measured inside the zone of remediation.

Demonstrating How the Technology Works

The second type of evidence needed to prove innovative remediation technologies—the cause-and-effect evidence—comes from data that link the basic risk reduction criteria with the technology being tested. The goal of collecting these data is to show that the physical, chemical, and biological characteristics of the site change in ways that are consistent with the processes initiated by the technology. Table 5-1 outlines, for each remediation technology subgroup identified in Chapter 3, the environmental conditions that can be monitored to establish the cause-and-effect linkage between remediation and the applied technology.

Carrying out many of the tests summarized in Table 5-1 will require the use of experimental controls. Experimental controls compare the differences in various site characteristics with and without application of the technology. The selection and use of controls in remediation technology testing are perhaps the most important factors in determining the success or failure of the experiment. Without good controls, it will be impossible to determine whether changes in site characteristics were a result of the technology application or of some other cause. Table 5-2 describes several control strategies that can be used to help determine which observed changes are a result of the remediation technology and which are not. Box 5-1 provides an example of experimental controls used to test a bioventing process.

The complexities of the subsurface and remediation technologies make computer models a useful tool for analyzing and generalizing results of remediation

TABLE 5-1   Data to Establish Cause-and-Effect Relationship Between
Technology and Remediation

Stabilization/Solidification/ Containment Technologies	Biological Reaction Technologies[a]

• Mechanism for decreased leachability
  - Formation of insoluble precipitate
  - Strong sorption/bonding to solids
  - Vitrification, cementing, encapsulation
• Integrity of stabilized material
  - Completeness of processes throughout treated region
  - Compressive strength of solidified material
  - Reaction to weathering (e.g., wet/dry and freeze/thaw tests)
  - Reaction to changes in ground water chemistry
  - Microstructural analyses of composition
• Geochemical conditions that affect leachability of stabilized materials (pH, Eh, competing ions, complexing agents, organic liquids, etc.)
• Increased ratio of immobile- to mobile-phase contaminants
• Fluid transport properties of solidified material
  - Permeability
  - Porosity
  - Hydraulic gradient across monolith
  - Rate of water flow through monolith
• Indicators of liquid/gas flow field consistent with technology (i.e., indication that flow through the stabilized or contained region is blocked)
	• Stoichiometry and mass balance between reactants and products
	• Increased concentrations of intermediate-stage and final products
	• Increased ratio of transformation product to reactant
	• Decreased ratio of reactant to inert tracer (or, in general, decreased ratio of transformable to nontransformable substances)
	• Increased ratio of transformation product to inert tracer (or, in general, increased ratio of transformation product to nontransformable substances)
	• Relative rates of transformation for different contaminants consistent with laboratory data
	• Increased number of bacteria in treatment zone
	• Increased number of protozoa in treatment zone
	• Increased inorganic carbon concentration
	• Changes in carbon isotope ratios (or, in general, in stable isotopes consistent with the biological process)
	• Decreased electron acceptor concentration
	• Increased rates of bacterial activity in treatment zone
	• Bacterial adaptation to contaminant in treatment zone
	• Indicators of liquid/gas flow field consistent with technology (i.e., indication that treatment fluids have been successfully delivered to the contaminated area)

[a]For further details about proving performance of biological reaction technologies, see National Research Council, 1993.

experiments. Whenever possible, computer simulation models should be used to plan and evaluate experiments to establish the link between observed remediation and the technology. Computer simulation models use mathematical equations to track the mass of contaminants in the subsurface. They describe how the contaminant mass is partitioned among aqueous and nonaqueous phases; how much is transported with the ground water, as a non aqueous-phase liquid (NAPL), or as a gas; how much reacts with other chemicals and with aquifer materials; how much degrades by biological or chemical reactions; and how each of these processes is

Chemical Reaction Technologies	Separation/Mobilization/ Extraction Technologies
• Stoichiometry and mass balance between reactants and products • Increased concentrations of transformation products • Increased concentrations of intermediate-stage products • Increased ratio of transformation product to reactant • Decreased ratio of reactant to inert tracer (or, in general, decreased ratio of transformable to nontransformable substances) • Increased ratio of transformation product to inert tracer (or, in general, increased ratio of transformation product to nontransformable substances) • Relative rates of transformation for different contaminants consistent with laboratory data • Changes in geochemical conditions, consistent with treatment reactions (pH, Eh, etc.) • Indicators of liquid/gas flow field consistent with technology (i.e., indication that treatment products have been successfully delivered to the contaminated material)	• Increased concentration (mass) of contaminant in outflow stream • Decreasing mass of contaminants remaining in subsurface consistent with mass extracted in outflow stream • Increased mass removal per unit volume of transport or carrier fluid • Increased ratio of contaminants in carrier fluid to aqueous-phase contaminants • Increased ratio of contaminants in carrier fluid to nonaqueous-phase contaminants • Observed movement of injected carrier fluids (flushing amendments or injected gases) or of tracers in carrier fluids • Spatial distribution of contaminants prior to, during, and after remediation • Indicators of liquid/gas flow field consistent with technology

affected by the introduction of a remediation technology. Simulations can be used in many ways in remediation technology evaluation. One approach is to use models to predict the behavior of contaminants under natural conditions and compare it with contaminant behavior during and following application of the remediation technology. A second approach is to use models to evaluate the sensitivity of soil or ground water quality variables to introduction of the remediation technology by simulating how those variables differ under natural and remediation conditions. A third approach is to use the model to quantify the uncertainty in various

TABLE 5-2 Experimental Controls for Improving Technology Evaluation

Method	Purpose
Collection of baseline data	Collection of accurate baseline data is the most basic type of experimental control and is essential to the success of the technology test. Without excellent baseline data, it will not be possible to develop an accurate comparison of conditions before and after application of the technology.
Controlled contaminant injection	In controlled contaminant injection, ground water from the site is spiked with the contaminants under consideration and re-injected into the aquifer. Therefore, the initial makeup, mass, location, and distribution of contaminants in the subsurface are known. Under these controlled conditions, the contaminant can be more easily and accurately tracked and monitored to determine the effect of the remediation technology.
Conservative tracers	Conservative tracers do not undergo the reactions associated with in situ reactive technologies. However, they are subject to a number of nonreactive processes that affect flow paths, flow rates, mixing, and retention of contaminants. Therefore, conservative tracers can be used to distinguish remediation resulting from the treatment process from that which occurs naturally.
Partitioning tracers	Partitioning tracers provide an indication of the total mass and spatial distribution of nonaqueous-phase liquids (NAPLs). They can be used to compare NAPL mass and spatial distribution prior to technology application with NAPL mass and distribution after remediation. Thus, they allow evaluation of NAPL removal and spatial patterns using a nondestructive technique.
Sequential start-and-stop testing	By alternating technology application and resting periods, the contaminant's fate can be observed under both natural conditions and remedial conditions. In this way the effects of the technology can be separated from remediation caused by naturally occurring processes. In addition, the start-and-stop approach can be used to distinguish between dynamic and equilibrium processes.
Side-by-side and sequential application of technologies	Side-by-side testing of two or more technologies at one site can be used to compare the capabilities of different technologies for the same hydrogeologic and contaminant setting. As an alternative, technologies can be applied sequentially at the same site to determine the marginal effectiveness of one technology over another.
Untreated controls	Untreated controls can help distinguish between technology-enhanced remediation and intrinsic remediation that occurs as a result of naturally occurring processes. The use of untreated controls is analogous to side-by-side testing with one of the remediation technologies being intrinsic remediation.
Systematic variation of technology's control parameters	The effect of changes in a technology's operating conditions on remediation can be determined by systematically changing control parameters. Ideally, this approach would be used to identify a technology's optimal operating conditions.

BOX 5-1
Use of Experimental Controls: Evaluating Bioventing at Hill and Tyndall Air Force Bases

At Hill Air Force Base in Utah and Tyndall Air Force Base in Florida, spills of JP-4 jet fuel have caused soil and ground water contamination. To demonstrate the capabilities of bioventing, the U.S. Air Force Center for Environmental Excellence sponsored field tests to evaluate the technology, which delivers oxygen to contaminated soils to stimulate contaminant biodegradation (Hinchee and Arthur, 1991; Hinchee et al., 1992).

Prior to the field tests at Hill and Tyndall, laboratory tests had shown that the addition of both moisture and nutrients may be needed to support continued contaminant biodegradation in bioventing systems. The field tests at both sites used experimental controls to quantify the effects of moisture and nutrient additions. At Hill, the bioventing system's parameters were sequentially varied to determine bioventing's effectiveness under different operating conditions. By operating the bioventing system first with no added moisture or nutrients, then adding moisture, then adding nutrients, researchers found that moisture addition stimulated biodegradation, but nutrient addition did not. At Tyndall, researchers used two side-by-side test cells to analyze the effects of moisture and nutrients. One cell received moisture and nutrients for the duration of the study. The other cell received neither moisture nor nutrients at the outset, then moisture only, then moisture and nutrients. In this case, no significant effect of either moisture or nutrients was observed. Researchers surmise that the different results at the two field sites were most likely due to contrasting climatic and hydrogeologic conditions. The fact that the two sites reacted differently indicates the need for additional controlled experiments to better gauge the effects of moisture and nutrients on bioventing.

---

types of data, allowing the user to evaluate the trade-offs between information, cost, and uncertainty when using different types of data. A final approach is to use models to determine the optimal experimental design to maximize information content of data while minimizing cost and uncertainty.

In proving that a technology is responsible for documented remediation and establishing the extent and rate of remediation attributable to the technology, a single type of evidence alone will usually not be sufficient. The larger the body of evidence used, and the more varied the converging lines of evidence, the stronger the case for the performance of the remediation technology.

Stabilization, Solidification, and Containment Technologies

When evaluating the performance of stabilization and solidification technologies, the most important data are those documenting immobilization of the contaminants. Thus, as indicated in Table 5-1, samples that document the mechanisms for decreased leachability (such as formation of an insoluble precipitate or cemented monolith) provide evidence that the stabilization technology has worked. Related to this will be data documenting the integrity of the stabilized material, such as data that demonstrate that the stabilization process is complete throughout the treated region or, for solidified material, data that document the permeability, porosity, and rate of fluid flow through the solidified monolith. Other data, such as the solidified material's compressive strength or its reaction to weathering tests, are an indication of the materials' long-term stability.

Stabilization, solidification, and containment technologies sometimes require certain environmental conditions to succeed. Properties such as pH, Eh, and concentrations of competing ions should be documented to show that geochemical conditions favor the stabilization processes at work. In addition, data can be collected to document changes in fluid flow fields that are consistent with the technology design.

Box 5-2 provides a case example of the types of data gathered to document performance of one type of solidification/stabilization process in a successful field test. This example provides a useful model for tests of solidification, stabilization, and containment technologies at other sites.

Biological and Chemical Reaction Technologies

In the process of transforming or immobilizing contaminants, biological and chemical reactions alter the soil and ground water chemistry in ways that can be documented to prove that the reaction processes are taking place. The observed chemical changes should follow directly from the chemical equations that define reactants and products and their ratios. Thus, many cause-and-effect data for biological and chemical reaction processes are derived from mass balance relationships defined by governing chemical equations. Increased concentrations of transformation products, concentrations of intermediate and final products, and ratios of reactants to products all can be used to demonstrate performance of biological and chemical reaction technologies. Geochemical conditions should also change in ways that can be predicted from the governing chemical equations. For example, ignoring microbial growth, the stoichiometric relationship used to relate oxygen (O_2) consumption and carbon dioxide (CO_2) production to biodegradation of petroleum hydrocarbons is

$$C_nH_m + (n + 0.25m)O_2 \rightarrow nCO_2 + (0.5m)H_2O$$

BOX 5-2
Proving In Situ Stabilization/Solidification of
Polychlorinated Biphenyls (PCBs) at the General Electric
Co. Electric Service Shop, Florida

International Waste Technologies (IWT)/Geo-Con conducted a field study to demonstrate the ability of their stabilization/solidification process to treat PCB-contaminated soils (EPA, 1990). The IWT in situ process mixes water and a cement-based proprietary additive with the contaminated soil to immobilize and contain PCB contaminants in a solidified, leach-resistant monolith. A series of analyses was performed on samples from the demonstration site to document stabilization/solidification of the PCBs in the soil. The table below describes the types of data that were collected to (1) document immobilization of PCBs and (2) establish the cause-and-effect relationship between the stabilization/solidification process and the documented remediation. A careful comparison of treated and untreated soils, along with a careful analysis of baseline conditions, provided the experimental controls for this evaluation.

Data Objective	Type of Data
Document PCB stabilization	• Leach tests showing immobilization of PCBs • Stabilized contaminant content of solidified soil
Link PCB immobilization to cementation	• Decrease in permeability of solidified material as compared to untreated soil • High unconfined compressive strength of solidified material • Documented integrity of solidified material under wet/dry weathering tests • Microstructural analyses—optical microscopy, scanning electron microscopy, and X-ray diffraction—showing that the solidified mass is dense, homogeneous, and of low porosity, with no compositional variations in the horizontal and vertical directions.

---

where C_nH_m represents a particular petroleum hydrocarbon. This equation can be used to determine how much O_2 will be consumed and how much CO_2 produced from the degradation of 1 mole (or 1 gram) of hydrocarbon or, conversely, how much hydrocarbon is degraded per mole of O_2 consumed or CO_2 produced. In other words, for every mole of O_2 consumed per minute, $1/(n + 0.25m)$ mole of hydrocarbon is degraded per minute; similarly, for every mole of CO_2 produced per minute, $1/n$ mole of hydrocarbon is degraded. Researchers from the U.S. Geological Survey (USGS) have used the above equation, along with field measurements of the rates of change of O_2 and CO_2 gas concentrations, to determine the rate of biodegradation of hydrocarbons at a site in Galloway Township, New Jersey (Lahvis and Baehr, 1996). The rates of biodegradation the USGS researchers computed based on O_2 consumption and on CO_2 production were in close agreement: the rate based on O_2 gas flux was 46.0 g per m^2 per year, while the rate based on CO_2 gas flux was 47.9 g per m^2 per year. The researchers used a mathematical transport model calibrated to the observed O_2 and CO_2 gas data to determine the O_2 and CO_2 gas fluxes and used a weighted average based on the concentrations of the various hydrocarbons found at the site to determine the stoichiometric coefficients.

Conservative tracers (see Table 5-2) are particularly useful when evaluating remediation systems that use biological or chemical reactions. Conservative tracers are not affected by biological and chemical reactions associated with the remediation technologies but are affected by all other nonreactive processes. Thus, they can be used to evaluate in situ reactions by documenting a decreased ratio of chemical reactant, or an increased ratio of transformation product, to tracer. For example, helium can be used in bioventing to show that O_2 loss is due to consumption by microorganisms rather than dispersion.

In recent years, considerable effort has focused on understanding the microbial reactions that degrade certain soil and ground water contaminants. In a 1993 report, the National Research Council outlined in detail the evidence required to document that bioremediation processes are occurring in the field (National Research Council, 1993). This evidence, in addition to using stoichiometric equations as described above, includes the number of bacteria, number of protozoa, rates of bacterial activity, and a range of other data that link observed ground water remediation with biodegradation processes.

Box 5-3 provides an example of data gathered to confirm a biological reaction process, in situ bioremediation of chlorinated solvents using methanotrophic bacteria. This example can serve as a model for other field tests of biological and chemical reaction processes.

Separation Technologies

Data collection for proving separation technologies should focus on documenting the transfer of contaminants to the more mobile liquid or gas phase and the level of increase in contaminant removal efficiency. The data should also show that the two (increased contaminant mobility and increased removal) are related. The most important piece of evidence is an increase in the concentration of contaminant in the fluid or gas extracted from the subsurface. This increase should coincide with a decrease in the mass of contamination remaining in the subsurface.

The success of separation technologies depends on delivering the carrier fluid to the subsurface contaminants. Thus, tracking the observed movement of injected fluids (such as flushing amendments in NAPL recovery or gases in air sparging) will be useful in linking contaminant mass transfer and removal with the arrival of the carrier fluids. Also, because many separation technologies involve altering the fluid flow field, documenting changes in fluid flow properties, such as fluid pressures and flow paths, that are consistent with the technology will be useful.

Boxes 5-4 and 5-5 provide examples of test protocols used to demonstrate two types of separation technologies (in situ mixed region vapor stripping and in situ cosolvent flushing) at two sites. These protocols can serve as models for future tests of separation technologies.

Determining the Level of Testing Required

The types of evaluations shown in Table 5-1 include theoretical modeling, laboratory experiments, and field tests. Which level of testing will be required depends on the complexity of the technology, but in general the strongest proof of technology performance derives from multiple lines of evidence demonstrating with laboratory and field data the theoretical concepts underlying the design of the technology.

Figure 5-1 shows how a technology would be proven under ideal circumstances: starting with theoretical concepts, proving these concepts in laboratory experiments and then in field tests, and then demonstrating the technology at full scale in the field. Some technologies, such as reactive barriers for in situ ground water remediation (see Box 5-6) have evolved through this linear, hierarchical process, from theory, to laboratory testing, to field testing, to full-scale application. For other commonly used technologies, however, the development process has been neither linear nor unidirectional. For example, early air sparging systems were designed based on field pilot tests, rather than detailed laboratory experiments, to prove efficacy and determine the extent of air flow, but the technology has matured based on careful laboratory studies to determine factors that influence the direction of air flow in the subsurface (see Box 5-7). SVE technology was applied in the field before detailed laboratory testing was conducted to fine-tune design procedures (see Box 3-2 in Chapter 3). Whether laboratory testing will be necessary before a technology is applied in the field depends on the complexity of the technology. For example, the design basis for SVE systems is fairly simple, involving induction of air flow to volatilize contaminants; detailed laboratory testing was not necessary prior to field testing because the processes controlling volatility are already well understood. On the other hand, the chemical reactions employed in reactive barriers are more complex and are highly sen-

BOX 5-3
Proving In Situ Bioremediation of Chlorinated Solvents at
Moffett Naval Air Station, California

Researchers at Stanford University conducted a field study to demonstrate engineered in situ bioremediation of chlorinated solvents by methane-oxidizing bacteria (Roberts et al., 1990; Semprini et al., 1990). In this experiment, known quantities of vinyl chloride, trichloroethylene (TCE), *cis*-dichloroethylene (*cis*-DCE), and *trans*-dichloroethylene (*trans*-DCE) were injected into a densely monitored, well-characterized aquifer. A series of biostimulation and bioremediation experiments was performed to document the engineered degradation of the organic solutes. Biostimulation by injection of methane- and oxygen-containing ground water was used to stimulate the growth of indigenous bacteria.

Results showed that biostimulation caused concurrent decreases in concentrations of the organic contaminants. The table below describes the types of data that were collected to (1) document remediation and (2) establish the cause-and-effect relationship between the methane-oxidizing bacteria and the documented remediation. In these experiments, controlled contaminant injections, conservative tracers, untreated test areas, systematic variation of operating parameters, and start-and-stop testing were used as controls.

Data Objective	Type of Data
Document reduction in quantity of contaminants	• Reduction in organic contaminant concentrations
	• Reduction in organic contaminant mass determined from the ratio of mean normalized concentration of organic contaminant to bromide tracer for quasi-steady-state conditions; a comparison of breakthrough of organic contaminants before and after biostimulation; and a mass balance comparing amounts of contaminant injected to amount removed at extraction wells
Link contaminant disappearance to indigenous methane-oxidizing bacteria	• Decrease of chlorinated organic contaminant concentrations coinciding with methane utilization.
	• Production of a transformation intermediate for *trans*-DCE

Data Objective	Type of Data
	• Increase in organic contaminant concentrations and disappearance of transformation intermediate when methane addition stopped
	• Relative transformation rates consistent with laboratory data (vinyl chloride degraded faster than *trans*-DCE, which degraded faster than *cis*-DCE, which degraded faster than TCE)
	• No degradation of TCE observed in zone where no methane was present to support bacterial growth
	• No evidence of anaerobic conditions (i.e., no intermediate products of anaerobic degradation)
	• Presence of indigenous methanotrophic bacteria

sitive to geochemistry; laboratory testing was essential in this case to define the parameters that control technology performance.

Whether or not a technology is laboratory tested prior to field application, the strongest proof of technology performance comes from multiple lines of evidence leading to the same conclusions. The evidence gathered should build a consistent, logical case that the technology works based on answering the questions of whether the risks from contamination decrease after application of the technology and whether the technology is responsible for the risk reduction achieved, as shown in the examples in Boxes 5-2, 5-3, 5-4, and 5-5.

Developing protocols that specify the general types of data that should be gathered for different technologies is possible, as shown in Table 5-1. However, the amount and specific types of data needed are highly specific to the individual technology and to the site where it is being tested. The data must minimize uncertainties associated with describing the complex heterogeneities of the subsurface environment, contaminant distribution, processes that control fate and transport of contaminants, and processes that control performance of the remediation technology. As a consequence, the details of the data collection plan and the extent of data collection vary for each new technology and each test of that technology in a new environment. The data collection plan should follow basic principles of experimental design, as outlined by Steinberg and Hunter (1984) and Cochran and Cox (1957). The data report should include a summary of the evaluation methods

BOX 5-4
Proving In Situ Mixed-Region Vapor Stripping in Low-Permeability Media at the Portsmouth Gaseous Diffusion Plant, Ohio

Researchers from Oak Ridge National Laboratory, Michigan Technological University, and Martin Marietta Energy Systems, Inc., conducted a full-scale field experiment to demonstrate the removal of volatile organic compounds (VOCs) from dense, low-permeability soils (West et al., 1995; Siegrist et al., 1995). The demonstration site, at a Department of Energy gaseous diffusion plant in Portsmouth, Ohio, had been used as a disposal site for waste oils and solvents. The silty clay deposits beneath the site were contaminated with VOCs at concentrations ranging up to 100 mg/kg. In addition, the shallow ground water was contaminated with trichloroethylene (TCE) at concentrations above the drinking water standard.

The remediation process, termed mixed-region vapor stripping (MRVS), mixes the soils in place using rotating augers. Compressed gases are injected into the mixed soils, and the VOCs are stripped from the subsurface. The off gases are captured at the surface and treated. The study included a set of replicated tests to evaluate the relative efficiencies of ambient and heated air for stripping VOCs. The following table describes the types of data that were collected to (1) document in situ stripping of VOCs from the dense, low-permeability layers and (2) establish the cause-and-effect relationship between the MRVS process and the documented remediation. A conservative tracer and systematic variation of operating parameters (heat) were used as controls.

used to prove technology performance similar to the examples in Boxes 5-2, 5-3, 5-4, and 5-5. The ranges of uncertainty for each type of data should be specified.

SELECTING A TEST SITE

In selecting a test site for an innovative remediation technology, technology developers usually confront one of two situations: either the developer will have a potential client and will need to demonstrate the technology at that client's site, or the developer will not have a client and will need to seek a test site available through various government programs. In the first case, the developer must face the question of how to select a location at the client's contaminated site to field test the technology. In the second case, the developer will need to apply to a government program to try to obtain a test site.

Data Objective	Type of Data
Document reduction in VOCs	• Reduction in soil VOC concentrations • Reduction in VOC mass in soils determined by analysis of off gases • Rate of VOC mass reduction determined from analysis of off gases
Link VOC reductions to in situ stripping	• TCE, 1,1,1-trichloroethane, and 1,1-dichloroethylene were present at same ratios in both off gas and soil matrix • Soil, air and off-gas temperature increased concurrent with injection of heated air • Absence of soil vapor pressure and temperature changes in undisturbed soil surrounding mixing zone, suggesting that VOCs in mixing zone were removed rather than being forced into surrounding soils • Tracer studies revealed that the process did not homogenize the soil and caused limited translocation of soil, suggesting that the VOCs in the mixing zone were removed rather than redistributed.

Testing at a Client's Site

For technology developers with an established client, the key to selecting a test location is to thoroughly characterize the contaminated site and then choose a test location that achieves a balance between being representative of conditions at the site and being simple enough that uncertainties in site hydrogeologic conditions do not overpower analysis and interpretation of technology performance data. There are four principal components to site characterization: (1) identification of contaminant sources, (2) delineation of site hydrogeology, (3) quantification of site geochemistry, and (4) evaluation of biogeochemical process dynamics. The types of data gathered for each of these components of site characterization will depend in part on the remediation technology being evaluated and in part on the types of contaminants present at the site. The end result of the site characterization will be a conceptual model showing locations of contaminant

BOX 5-5
Proving In Situ Cosolvent Flushing at Hill Air Force Base

At a Hill Air Force Base site contaminated with jet fuels and chlorinated solvents disposed of in the 1940s and 1950s, University of Florida and EPA researchers conducted a pilot-scale field study to demonstrate enhanced contaminant solubilization by in situ cosolvent flushing (Annable et al., in press; Rao et al., in review). The researchers installed a test cell in a 2-m-thick contaminant source zone containing a large mass of contaminants present as NAPLs. The test cell dimensions were 5 m x 3 m x 10 m deep. The cell was underlain by a deep clay confining unit, so that the test zone was hydraulically isolated from the rest of the aquifer.

Over a 10-day period, the researchers injected a total of 40,000 liters of a cosolvent mixture (70 percent ethanol, 12 percent pentanol, and 18 percent water) through four injection wells. Following the cosolvent injection, the researchers flushed the test cell extensively with water to remove the cosolvents. The cosolvent fluids, along with solubilized NAPL, were extracted through three wells. A network of 72 multilevel samplers allowed monitoring of the internal dynamics of the flushing process between the injection and extraction wells.

The researchers collected a variety of data, shown in the table below, to establish multiple line of evidence for NAPL removal and to link NAPL removal with the cosolvent flushing process. As predicted in earlier laboratory studies, cosolvent flushing in the field test removed more than 95 percent of several NAPL constituents and more than 85 percent of total NAPL mass. The unextracted NAPL mass was highly insoluble and contained no measurable concentrations of target contaminants. Sequential applications of a pump-and-treat system and the cosolvent flushing system, as well as comparisons of the movement of conservative tracers and partitioning tracers before and after remediation, were used as experimental controls.

Data Objective	Type of Data
Document reductions in NAPL mass and contaminant concentrations	• Decreased concentrations of NAPL constituents in soil cores • Decreases in NAPL constituent concentrations in ground water samples • Increased concentrations of NAPL constituents in extraction fluids • Decreased retardation of partitioning tracers after treatment, indicating extraction of NAPL mass

Data Objective	Type of Data
Link NAPL removal to cosolvent flushing	• Consistency of mass removal estimates from all of the above evaluations; all showed greater than 85 percent removal of NAPL mass • Monitoring of contaminants, cosolvents, and tracers outside the test cell, demonstrating the effectiveness of hydraulic containment • Comparison of NAPL removal achieved with a conventional pump-and-treat system and that achieved with cosolvent flushing; extensive flushing with a pump-and-treat system did not lead to decreased contaminant concentrations in produced fluids • Large rise in dissolved NAPL concentrations coincident with arrival of cosolvents in samples taken at the multilevel monitoring wells and extraction wells • Maximum NAPL constituent concentrations in extracted fluids consistent with predictions based on controlled laboratory studies of NAPL solubilization with the cosolvent mixtures.

sources and plumes and site hydrogeology (see Figure 5-2), along with tables showing site biogeochemistry and important biogeochemical processes. The data collected during site characterization must be sufficient to provide a baseline for assessing (1) whether the technology works and (2) how it works.

The first component of site characterization—the identification of contaminant sources—includes evaluation of the types of contaminants in the subsurface and their properties (reactivity, solubility, volatility, and mobility). It also includes delineation of the spatial distribution and measurement of the concentrations of contaminants in the subsurface, with particular attention to the distribution of contaminants among the aqueous, nonaqueous, solid, and sorbed phases. This step of site characterization will be the same regardless of the type of technology being tested, because a thorough documentation of contaminant distribution is essential for designing the technology installation and understanding the

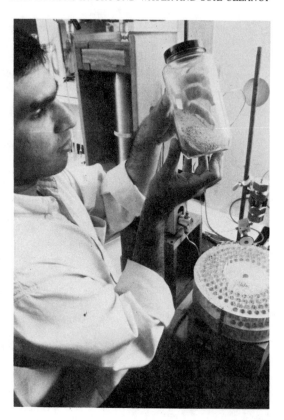

A column experiment to evaluate the kinetics and chemistry of surfactant flushing of hydrophobic organic contaminants from aquifer sediment. Courtesy of Richard Luthy, Carnegie Mellon University.

technology's performance. The disposal or spill history at the site and site hydrogeology play important roles in determining the distributions of contaminant sources and the sizes and shapes of plumes generated from these sources. Since detailed historical records are often not maintained, delineation of contaminant sources is the most challenging problem in site evaluation. Especially difficult to map are NAPL sources, because conventional soil coring methods cannot provide a complete picture of NAPL distribution. However, recently developed tracer techniques offer considerable promise for mapping NAPL mass distribution using nondestructive techniques (see Box 5-8 for details).

The second component of site characterization—delineation of site hydrogeology—involves developing a model of site geologic units and quantifying hydrogeologic properties that influence ground water and contaminant movement. This step in site characterization will be the same regardless of the type of remediation technology being tested, because a detailed understanding of ground water and contaminant movement is essential for designing pilot tests of any remediation technology. Included in this stage of site characterization are an

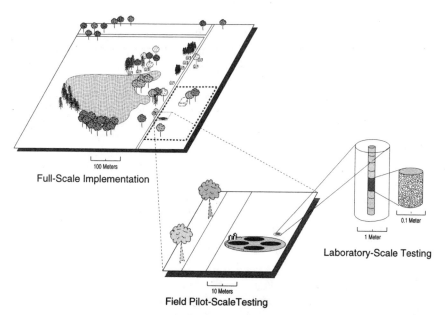

Full-Scale Implementation

Laboratory-Scale Testing

Field Pilot-ScaleTesting

FIGURE 5-1   Usual hierarchical scales of technology testing. Laboratory-scale testing is used to identify and quantify factors that affect process performance. Field testing is used to determine whether the technology can work under real-world conditions and to modify designs based on laboratory results. Full-scale implementation is needed before a final verdict can be reached on the effectiveness of a technology. SOURCE: Modified, with permission, from Benson and Scaife (1987). © 1987 by A. A. Balkema.

evaluation of site geology; characterization of stratigraphy, including types, thicknesses, and lateral extent of aquifer units and confining units; measurement of depth to ground water, ground water recharge and discharge points, hydraulic gradients, and preferential flow paths; quantification of aquifer physical properties, including hydraulic conductivity, porosity, and grain size distribution, as well as the variations in these properties with location; and quantification of vadose zone properties, including gas and water permeability.

The third component of site characterization—quantification of aquifer geochemistry—involves measuring any chemical properties of the ground water that will affect the fate of the contaminants and performance of the remediation technology. Thus, if technology performance is sensitive to changes in pH, Eh, dissolved oxygen concentration, organic carbon concentration, or any other geochemical parameter, then these parameters must be carefully documented before remediation begins.

The fourth component of site characterization—evaluation of biogeochemical process dynamics—includes analyzing geochemical characteristics of the soil

---

BOX 5-6
Development of Passive-Reactive Barriers Based on
Laboratory Studies

The concept that zero-valent metals, such as iron, can dehalogenate chlorinated compounds and thus might be useful in environmental cleanup first appeared in the scientific literature in the early 1970s. However, this concept was not extended to the cleanup of contaminated ground water until approximately 1990, when researchers at the University of Waterloo began laboratory and field studies to determine whether zero-valent metals might be applicable to cleanup of contaminated ground water (Gillham, 1995). Their idea was to emplace zero-valent iron in a permeable underground wall ahead of a plume of contaminated ground water, so that any chlorinated compounds in water passing through the wall would be dechlorinated by the iron (see Chapter 3) (Gillham et al., 1994).

In laboratory batch and column experiments designed to mimic the ground water environment and in a field study, the Waterloo researchers demonstrated chlorine mass balances of 100 percent, showing that the contaminants were dechlorinated by zero-valent iron (Orth and Gillham, 1996). Early testing also included determination of dechlorination reaction rates in laboratory studies with a number of halogenated methanes, ethanes, and ethenes to demonstrate the potential applicability of this method to a range of contaminants and to document reaction requirements (Gillham and O'Hannesin, 1994). Key experiments showed that metallic iron creates low redox conditions necessary for the dechlorination of the chlorinated compounds and that iron solid is needed for the reaction to proceed. These experiments have led to several full-scale applications of the technology (see Box 3-5 in Chapter 3 for one example).

Prior to installation of a permeable, iron-containing reactive wall, treat-

---

and aquifer materials, mineralogy, sorption potential of solid materials, presence or absence of indigenous microbes and their biodegradation potential, nutrient conditions, substances that may inhibit or compete with biodegradation, and any other biogeochemical properties that might play a role in remediation and in the natural fate of contaminants in the absence of remediation. Design of a data collection plan for understanding biogeochemical process dynamics will vary with the type of remediation technology being tested, because different types of technologies will be influenced by different biogeochemical processes. However, for all technologies, enough data on these processes should be gathered to allow an understanding of the fate of contaminants in the absence of remediation so that

ability studies are used to identify design parameters for the wall. Treatability studies for the first full-scale applications of the technology were expensive, consisting of laboratory column experiments using site ground water, variable mixtures of sand and iron, and varying water velocities, followed by similar studies in the field using above-ground canisters. As experience and acceptance of the technology have increased, the treatability study protocol has been streamlined to consist generally of column experiments with 100 percent reactive iron and ground water from the site (ETI, 1995). While the testing and design phase prior to the first full-scale installation (see Box 3-5) required nearly three years, application at a second site in the same state required only a few months of testing. In some instances, for ground water with low contaminant concentrations, mixtures containing half sand and half reactive iron are tested. The laboratory columns are monitored for contaminants and reaction products with time until reaching steady state. Pseudo-first-order rate coefficients for each contaminant are determined from the steady-state concentration distributions. System designers then determine the required residence time in, and thus the thickness of, the reactive wall based on these rate coefficients (Thomas et al., 1995). At large sites, pilot-scale studies may also be needed to modify and improve the design. In all of these studies, temperature and pH of the actual field conditions must be mimicked, because these parameters can significantly affect transformation rates. Designers must also characterize the site's hydrogeologic conditions adequately to ensure that the reactive wall will capture the contaminant plume.

The relatively rapid commercialization of passive reactive barriers was due in part to the well-planned laboratory and pilot tests preceding the first commercial application. Research at independent laboratories provided independent confirmation of rapid contaminant transformation rates, and the well-planned pilot test demonstrated success in the field.

the effect of remediation can be assessed. That is, the data must allow a determination of the extent to which any reductions in contaminant concentration and mass can be attributed to intrinsic biogeochemical processes that occur in the absence of remediation. Estimating biogeochemical process parameters is an uncertain exercise at many sites. The parameters may vary both spatially and over time due to microscale variations in environmental and geochemical properties.

Once the site is adequately characterized, a test plot should be chosen that represents conditions at the site but is simple enough to minimize uncertainties in evaluating technology performance. Ensuring that the test plot is reasonably representative of the site is essential for the scaleup stage of the technology test. If

Laboratory-scale testing to evaluate the transport of microorganisms used for biological treatment in an aquifer. Courtesy of Roger Olsen, Camp Dresser & Mckee.

the pilot test site is not geostatistically representative of the section of the site for which the technology is being considered, then the technology may fail during scaleup. At the same time, in the early stages of technology development, more useful information is gained by conducting tests in relatively uncomplicated geologic settings, which allow the developer to separate inherent process performance limitations from matrix complexities. Unfortunately, such simple sites are not always available, resulting in a data set that is confounded by geologic complexity. This is especially an issue for sites where the technology "fails" its demonstration; it may be unclear whether the failure is inherent to the remediation process or is due to complex site conditions and inadequate accounting for these conditions in the design of the remediation system.

While the test area must be simple enough to allow evaluation of technology performance, at the same time it must be representative of site conditions. That is, the test volume must contain a geostatistically representative number of the geologic and contaminant features likely to be critical in full-scale project implementation. Otherwise, the uncertainty in extrapolating results from the test cell to full-scale application will be too large to allow for meaningful predictions. Failure to select a representative test area during a pilot test can often lead to unanticipated technical difficulties, less effective remediation than indicated in the pilot tests, and cost overruns when the full-scale project is initiated.

Pilot-scale land biotreatment units used to assess biotransformation and biodegradation of PCB-contaminated sediment and sludge. Pilot-scale tests are needed to assess practical treatment rates. Courtesy of Alcoa.

A final consideration in the selection of a test location is concerns of site regulators. Obtaining regulatory approval to test technologies involving injection of substances, either treatment fluids or contaminants to be used in the test itself, can be a complex process (see Boxes 5-9 and 5-10). For example, in the case described in Box 5-10, involving selection of a site to conduct tests of multiple technologies, selecting a test site and obtaining all of the necessary regulatory permits took one-and-a-half years, much longer than project managers and the consortium of academic researchers involved in the study had anticipated.

Testing at a National Test Site

Until the mid-1990s, very few sites were available for researchers to test innovative remediation technologies in the field without first having a client interested in buying the technology. Essentially the only option was to test technologies under the Superfund Innovative Technology Evaluation (SITE) Program, which has limited scope and funding. (In fact, funding for the SITE Program was

BOX 5-7
Development of Air Sparging Based on Field Applications
Followed by Detailed Studies

As with many remediation technologies, the initial application of air sparging was based on a rudimentary understanding of the technology and was field derived. Air sparging design was based primarily on field pilot testing to prove efficacy and determine the extent of air flow. However, air sparging has rapidly progressed from an empirical field practice to an ongoing research area.

Designers of air sparging systems have benefited from the lessons learned from the development of SVE: that air flow is the key to successful treatment and that air flow in the subsurface is governed by a number of parameters. Early protocols for testing air sparging systems used monitoring well parameters such as pressure readings, dissolved oxygen concentrations, contaminant volatilization rates, and water table rise as indicators of the effective radius of the sparging systems. While these parameters provide a general indication of where the injected air is traveling, smaller sampling intervals and the use of tracer gases have shown that air flow is complex, not predictable, and not as uniform as would be indicated by monitoring well data. In fact, recent studies have shown that even with conventional pilot testing and design, air sparging performance can be highly variable. Johnson et al. (1997) conducted pilot and full-scale tests of air sparging in a gasoline-contaminated shallow aquifer and concluded that short-term pilot tests involving measurement of the typical parameters (dissolved oxygen concentrations, pressure readings, water table levels) used to estimate air sparging performance provided overly optimistic estimates of long-term, full-scale system performance. Based on such findings, Johnson and others have recommended that short-term pilot tests of air sparging be used to evaluate the feasibility of using the technology and to identify reasons for poor performance under test conditions, rather than to provide detailed predictions of long-term performance (Johnson et al., 1997).

Current air sparging research is focusing on improving methods for measuring and controlling air flow in saturated media. If measurement tools can be developed to determine where air travels during sparging, design of systems and predictions of performance can be more rigorous. Methods for maximizing air distribution during sparging include pulsing and multilevel sparging.

eliminated in 1996 and reinstated in 1997 at a modest $6 million, a fraction of the cost of cleaning up one Superfund site.) In addition, only one site, the Moffett Air Force Base in California, was available for researchers to inject contaminants under controlled conditions and monitor the results of a remediation system (National Research Council, 1994).

In recent years, the federal government has made a significant effort to increase the number of test sites available on federal facilities, including sites where controlled contaminant injections are allowed for technology evaluation purposes. In 1996, the EPA issued a policy memorandum providing strong encouragement to use federal facilities as sites for demonstrating technologies (Laws, 1996). Table 5-3 lists federal programs that provide assistance to developers needing sites or other forms of support for field testing innovative remediation technologies. Some of these programs, such as the Navy's Environmental Leadership Program, guarantee full-scale use of any technology successfully demonstrated under the program (EPA, 1996a). Such programs provide critical support for innovative remediation technology developers at the stage between technology development and commercialization.

While the number of test sites has increased, competition for pursuing tests under these federal programs is intense. For example, the first solicitation for technology testing to be carried out at Department of Defense sites under the Advanced Applied Technology Demonstration Facility program drew 170 proposals; only 12 of these were selected (see Box 5-11).

SITE-TO-SITE TRANSFER OF TECHNOLOGIES

Once a technology has been successfully tested at one site, the developer and potential clients will wish to determine whether the technology can be transferred to another site. In general, some degree of additional testing at the second site will be required prior to implementing the technology there. However, funds for field pilot testing are often limited, and the technology developer is often faced with the task of minimizing the amount of additional site-specific testing, while at the same time providing adequate assurance that the technology will perform as predicted at the new site. An adequate understanding of how the technology works (that is, an answer to the second question posed at the beginning of this chapter) can help to minimize site-specific pilot testing costs.

The degree of additional testing that will be required before an innovative remediation technology can be applied at the second site is primarily a function of the properties of the contaminant and the hydrogeologic setting in which it exists. In general, technologies used to treat mobile and/or reactive contaminants will require less additional testing than those used to treat contaminants with limited mobility and reactivity. Solubility and volatility are the primary factors that control contaminant mobility in soil and ground water. Reactivity is a measure of the biodegradability or chemical reactivity (via oxidation, reduction, or

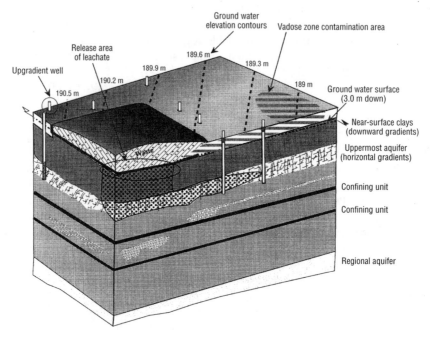

FIGURE 5-2   Development of a conceptual model of the geologic units of a contaminated site.

precipitation) of the contaminant. The more a treatment technology makes use of a fundamental property related to contaminant mobility or reactivity, the more easily it can be transferred from one site to another for treatment of similar contaminants. Similarly, the more conducive a geologic setting is to fluid (air and/or water) flow, the more easily a new technology can be applied at that site with minimal additional testing. Permeability and degree of saturation are the two hydrogeologic factors that most affect treatability.

As discussed in this chapter, there are generally two purposes for testing the performance of innovative remediation technologies. One is to prove the efficacy of a process: Does it reduce the risks posed by the soil and/or ground water contamination? The other is to determine how the process works: Which contaminant properties does the technology make use of, and how is the process affected by hydrogeologic properties such as permeability and saturation? The higher the treatability of a contaminated site, the lower the site-specific testing requirements will be for assessing a new technology's efficacy or applicability. For example, application of SVE at a site with highly volatile contaminants in a highly permeable formation would require limited site-specific testing. On the other hand,

BOX 5-8
NAPL Source Zone Mapping: Use of Tracer Techniques

Locating, quantifying, and delineating NAPL source zones presents considerable difficulties and uncertainties due to the highly heterogeneous nature of NAPL distribution in subsurface zones. Identifying the nature and extent of NAPL source contamination is an essential element of site characterization and a regulatory requirement. Conventional geophysical methods used for NAPL source mapping include soil core sampling, ground water and soil gas analyses, electromagnetic resistivity tests, and ground penetrating radar techniques (Feenstra and Cherry, 1996). The most common of these are analyses of samples (soil, gas, or ground water) taken at several locations at a site. These point measurements of NAPL contaminant concentrations are spatially interpolated to estimate total NAPL mass. However, such estimates are subject to considerable uncertainties. Also, some of these techniques require destructive sampling (as in soil coring), precluding repetitive sampling at the same location.

A new experimental technique, based on the displacement of a suite of tracers through the NAPL source zone, was developed by researchers at the University of Texas (Jin et al., 1995). Data from nonreactive tracers yield information about site hydrodynamic characteristics, and the extent of retarded transport of the reactive tracers yields a measure of the NAPL volume present in the zone swept by the tracers. Multilevel samplers placed between the injection and extraction wells provide data to map the NAPL spatial distribution, and the extraction wells provide depth- and volume-integrated estimates of total NAPL volume.

The first field-scale test of the partitioning tracer technique was conducted by University of Florida researchers at Hill Air Force Base (Annable et al., in press). This test provided an integrated measure of the total volume of NAPL and its spatial distribution within an isolated test cell (3 m × 5 m × 10 m), and the tracer results were consistent with average values estimated from soil core and ground water analyses. Similar tracer tests were completed in eight other test cells as part of a coordinated study at the same site; results are expected to be released by the end of 1997.

The use of this tracer technology is in its early stages of development. As additional data from a variety of test sites are gathered, use of the technique will increase.

BOX 5-9
State Regulatory Policies for Remediation Technology Testing

For remediation technology developers, obtaining appropriate regulatory approvals for technology testing is a major hurdle that must be overcome in order to have a chance to demonstrate the new technology under field conditions. This is an especially significant problems for in situ flushing technologies that require injection (and withdrawal) of additives such as surfactants and alcohols for enhanced extraction and for some remediation technologies requiring addition of nutrients, primary substrates, or electron acceptors. The underground injection of additives is prohibited by regulatory or procedural barriers in many states. Authority for regulation of underground injection wells is split between the states and the federal government. The EPA conducted a survey in 1995 to identify institutional barriers to remediation technologies that require some type of underground injection. The report (EPA, 1996c) reached the following conclusions:

• About two-thirds of the states have allowed some sort of injection incidental to an in situ ground water remediation technology; most of these cases were for injection of nutrients for enhanced bioremediation.
• Eleven states have allowed surfactant injection, mostly at Superfund sites. (One state has allowed alcohol injection since this EPA report was published; see Box 5-5 for the case study.)
• No state has a direct regulatory prohibition of injection technologies for treating contaminated aquifers. A few states have policies that discourage use of injection technologies; however, most of the states have rejected most or all of the proposals received, citing a broad spectrum of concerns.
• The technical merit of the proposed technology, as documented in the proposal to the state, was the key factor in the approval process.

---

application of a process claiming to biodegrade DDT would require significant testing.

Figure 5-3 and Tables 5-4 and 5-5 depict factors influencing treatability and show how contaminant properties and geologic setting affect treatability and the need for site-specific testing of remediation technologies. The columns in Figure 5-3 organize contaminants on the basis of high (H) or low (L) volatility, reactivity, and solubility. The rows organize geologic settings on the basis of texture and saturation. The figure shows four groups of contaminated sites: easy to treat (category I), moderately difficult to treat (category II), difficult to treat (category III),

Installing electrodes and treatment zones for a pilot test of electro-osmosis treatment at the Paducah, Kentucky, Department of Energy site. The pilot test was conducted by the Lasagna Consortium™. Courtesy of the Department of Energy.

and extremely difficult to treat with current technologies (category IV). As the difficulty of treatment increases, so does the amount of site-specific testing required to prove efficacy and applicability.

Category I includes sites with highly volatile and/or reactive contaminants in highly permeable soils. Contaminated sites in this category are easy to treat primarily because volatilization or biodegradation can efficiently remove contaminant mass. Contaminants in this category include the gasoline components benzene, toluene, ethylbenzene, and xylene (BTEX); chlorinated volatile organics such as chlorinated ethenes; and alcohols. High contaminant solubility complicates in situ treatment because it causes more contaminant mass to dissolve in ground water, making contaminant volatilization more difficult. However, the high volatility and reactivity of contaminants in category I make treatment of these contaminants in homogeneous saturated aquifer formations relatively easy.

For sites in category I, determining the efficacy and applicability of a remediation process generally requires minimal site-specific testing, as shown in Table 5-5. Efficacy and applicability of a technology can often be determined from the fundamental properties of the contaminant or the site. Testing requirements for remediation technologies being considered at sites in this category are generally

---

BOX 5-10
Selecting a Test Site for Side-by-Side Technology
Comparisons

The Strategic Environmental Research and Development Program (SERDP), funded by the Departments of Defense (DOD) and Energy and the EPA, has funded a program to evaluate performance of several innovative remediation technologies side by side. The first step in the program was to identify a portion of a field site where innovative technologies could be pilot tested side by side. The program stipulates that the tests be conducted at a DOD facility unless a suitable DOD site cannot be found. Despite a high level of cooperation from DOD site managers and regulatory officials, selecting a test site and obtaining all of the necessary regulatory permits took a year and a half, much longer than project managers had anticipated (C. Enfield, EPA, personal communication, 1995).

The criteria used for identifying an "ideal" test site for this project included the following: shallow ground water with a confining layer less than 15 m below ground surface in order to minimize testing costs and concern about off-site impacts; a permeable aquifer to ensure that tests can be conducted in a relatively short period of time; presence of a single-component NAPL, preferably a dense NAPL (DNAPL), as residual saturation or in pools; a NAPL source area large enough to accommodate multiple test cells; a secure site with convenient access and infrastructure support; cooperative site owners; and a flexible regulatory permitting process. In the early stages of the project, most candidate sites were eliminated from consideration due to one or more of the following problems: regulatory constraints and liability concerns; inability to locate the contaminant source area with certainty; inadequate size of the source area; presence of multiple sources of contamination or complex wastes

---

dictated more by site-specific design requirements than by questions of efficacy or applicability. For example, SVE technology could be assumed to be an effective remedy for treatment of trichloroethylene in a medium-grained sand, but application of SVE to such a site would require pilot testing to optimize the technology's design parameters. A technology for treating a site in this category is easily extended to other sites within this same category with contaminants for which the technology is appropriate.

Category II represents contaminated sites that are moderately difficult to treat. Examples include sites with contaminants having low volatility and high solubility in geologic settings having high to moderate permeability (as in column D in Figure 5-3). Many of the contaminants in this category are also either biologically or chemically reactive; examples include phenols, glycols, methyl tertiary-butyl

with multiple components; and inadequate infrastructure and support services, access, or security.

The site that was ultimately selected for testing was Operable Unit 1 at Utah's Hill Air Force Base, a Superfund site. The site meets many of the criteria listed above. The Air Force is the only party liable for site contamination. The shallow, unconfined, sand-and-gravel aquifer is underlain by a thick, confining clay unit about 10 to 15 m below ground surface that extends for several hundred meters. The water table is located at about 8 m below ground surface. The aquifer is contaminated with a complex NAPL consisting of a mixture of aviation fuel (JP-4), waste solvents disposed of in two chemical disposal pits during the 1940s and 1950s, and fuels and combustion products from a fire training area. Contaminants targeted for remediation include aromatic petroleum hydrocarbons and n-alkanes from the aviation fuel and chlorinated alkenes and chlorobenzenes from the solvents. The NAPL source area and the associated dissolved plume cover an area of about 8 ha. Residual NAPL is present as a 2-m-thick smear zone just above the clay unit.

Remediation technology testing was carried out in nine test cells, each 3 m × 5 m. The test cells are hydraulically isolated from the rest of the aquifer. Isolation was achieved by driving interlocking sheet piles keyed into the underlying clay confining unit and sealing all the joints with grout (Starr et al., 1993). All are instrumented in essentially the same manner, with four fully screened injection wells, three fully screened extraction wells, and multilevel samplers in at least 72 locations (on a 0.7 m × 0.7 m grid). The distribution of NAPL in each cell was carefully characterized by soil coring and partitioning tracer tests prior to testing. The remediation technologies tested in the cells are several varieties of in situ methods for NAPL extraction; the methods use either cosolvents, surfactants, steam, air, or cyclodextrin to extract the NAPLs. Enhanced solubilization and enhanced mobilization are the two methods for NAPL extraction.

ether, and naphthalenes or other polycyclic aromatic hydrocarbons having three or fewer benzene rings.

When being considered for use in treating sites in category II, innovative remediation technologies will require a moderate amount of testing, as shown in Table 5-5. Assurance of performance may be strongly indicated by fundamental properties of the contaminant (such as biodegradability) or the site (such as transmissivity to fluids), but performance cannot be predicted with certainty based on these properties. Testing is required to determine applicability, especially if chemical or biological reactivity is the basis for treatment. Testing is usually directed at identifying conditions that may limit the applicability of the technology to the site. For example, application of a bioremediation process at a site contaminated with phenol would require testing to determine that the site geo-

TABLE 5-3   Federal Programs Providing Support to Remediation Technology Developers

Program	Purpose	Contacts
National Environmental Technology Test Sites (NETTS) Program	Provides locations, facilities, and support for applied research, demonstration, and evaluation of innovative subsurface cleanup and characterization technologies that are candidates for restoration of Department of Defense (DOD) facilities	Dr. Mark Noll, Air Force (302) 678-8284 Ernest Lory, Navy (805) 982-1299
Advanced Applied Technology Demonstration Facility (AATDF)	Seeks to identify, demonstrate, and commercialize advanced technologies potentially useful in ground water and soil remediation at DOD facilities	Dr. Herb Ward, Rice University (713) 527-4725
Rapid Commercialization Initiative (RCI)	Provides in-kind assistance for selected companies with commercially ready environmental technologies that require demonstration and performance verification	Stanley Chanesman, Department of Commerce (202) 482-0825
Strategic Environmental Research and Development Program (SERDP)	Seeks to identify, develop, demonstrate, and implement technologies of use to the DOD in six areas, including environmental cleanup	Dr. Olufemi Ayorinde, DOD (703) 696-2118
Wurtsmith Air Force Base National Center for Integrated Bioremediation Research and Development (NCIBRD)	Allows testing of biological and other technologies for remediation of fuels and solvents; tests are conducted at Wurtsmith Air Force Base	Dr. Michael Barcelona, University of Michigan (313) 763-6512
Air Force Center for Environmental Excellence (AFCEE) Innovative Technology Program	Identifies and field tests innovative environmental technologies, including remediation technologies	John Caporal, Air Force (210) 536-2394
Environmental Security Technology Certification Program (ESTCP)	Selects laboratory-proven technologies with DOD market application and moves them to the field for rigorous testing	Dr. Jeffrey Marqusee, DOD (703) 614-3090

TABLE 5-3 Continued

Program	Purpose	Contacts
Naval Environmental Leadership Program (NELP)	Selects innovative remediation technologies for full-scale application at Naval Air Station North Island in San Diego, California, and Naval Station Mayport in Jacksonville, Florida	Ted Zagrobelny, Navy (703) 325-8176
Superfund Innovative Technology Evaluation (SITE) Program	Supports bench- and pilot-scale studies of innovative remediation technologies	Annette Gatchette, EPA (513) 569-7697

SOURCE: Adapted from EPA, 1996a.

chemical conditions (availability of nutrients and oxygen) required for effective performance are present. For technologies being considered for sites in this category, the development of a data base showing all prior technology applications is essential to commercialization, and development of a common data collection and reporting protocol would greatly assist in expanding use of the technologies. As the data base grows, the need for site-specific testing would diminish.

Category III represents contaminated sites at which the contaminant is soluble but is neither reactive nor volatile and/or at which the geology is heterogeneous. Many sites contaminated with inorganic chemicals are in this category. As shown in Table 5-5, neither the efficacy nor the applicability of technologies for treating such sites is easily derived from the fundamental properties of the contaminant or the site. Characterizing the hydrologic and geochemical variability of the site and the influence of hydrologic and geochemical properties on contaminant retention and reaction processes is extremely difficult for category III sites. Testing at each individual site is required to prove efficacy and to determine applicability. Testing may have a number of stages, including laboratory, pilot, and full scale, but results can be readily transferred from one stage of testing to another. As testing progresses from one stage to another, the focus changes from proof of efficacy or applicability to site-specific design.

The final category in Figure 5-3 represents sites with contaminants that are neither volatile, nor reactive, nor soluble and/or having complex geologies such as clay and fractured rock. Contaminants in this category include polychlorinated biphenyls, organochlorine and organophosphorus pesticides, and polycyclic aromatic hydrocarbons with more than three benzene rings. Such contaminants are extremely difficult to treat with existing commercial technologies because their low solubility and volatility and high sorption potential complicate their detection, analysis, and destruction or removal from the subsurface. Treatment of sites

---

BOX 5-11
Technology Testing Under the Advanced Applied
Technology Demonstration Facility Project

In 1993, the Department of Defense (DOD) initiated a program known as the Advanced Applied Technology Demonstration Facility (AATDF) project. The DOD budgeted $19.3 million for the project to support field testing of innovative technologies for characterization and cleanup of contaminated ground water and soil. The program is administered by a university consortium including Rice University (the lead institution), Stanford University, the University of Texas, Lamar University, Louisiana State University, and the University of Waterloo. It is supported by five major consulting firms and an advisory group including representatives from DOD and industry.

Competition for obtaining funds to support technology testing under the AATDF program has been intense. The initial solicitation yielded 170 proposals; 38 of these were selected for submission of full proposals, and 12 of these 38 were selected for funding. Funded projects include field testing of funnel-and-gate technologies for directing ground water flow into a treatment zone, soil fracturing and steam injection for treatment of semivolatile contaminants in low-permeability zones, phytoremediation and mining technologies for removing lead contamination from soil, in situ cooxidation technologies for treating trichloroethylene and jet fuel, radio frequency heating for improving removal of semivolatile compounds, surfactant injection for treatment of NAPLs, and single-phase microemulsion treatment for removal of NAPLs. Also funded in this first round of AATDF projects are an investigation of a laser fluorescence cone penetrometer method for site characterization, development of technical practices manuals for successfully demonstrated technologies, and development of an experimental controlled release site where technologies can be tested following controlled releases of contaminants. The 12 projects are due to be completed by the end of 1997.

---

in complex geologic settings is difficult to assess because of the difficulty of obtaining representative data. Detailed laboratory, pilot, and field tests are fundamental to proving either efficacy or applicability of new technologies designed to restore these types of sites. A critical question in the development of technologies for this category is how easily data may be extrapolated from one stage of testing to the next due to the difficulty of obtaining data and the inherent variability of the data. For example, determining what size of pilot test area is necessary to adequately represent the full site may be difficult. As shown in Table 5-5, multiple pilot tests may be necessary. A problem in determining either the efficacy or the applicability of technologies for sites in this category is that success at one

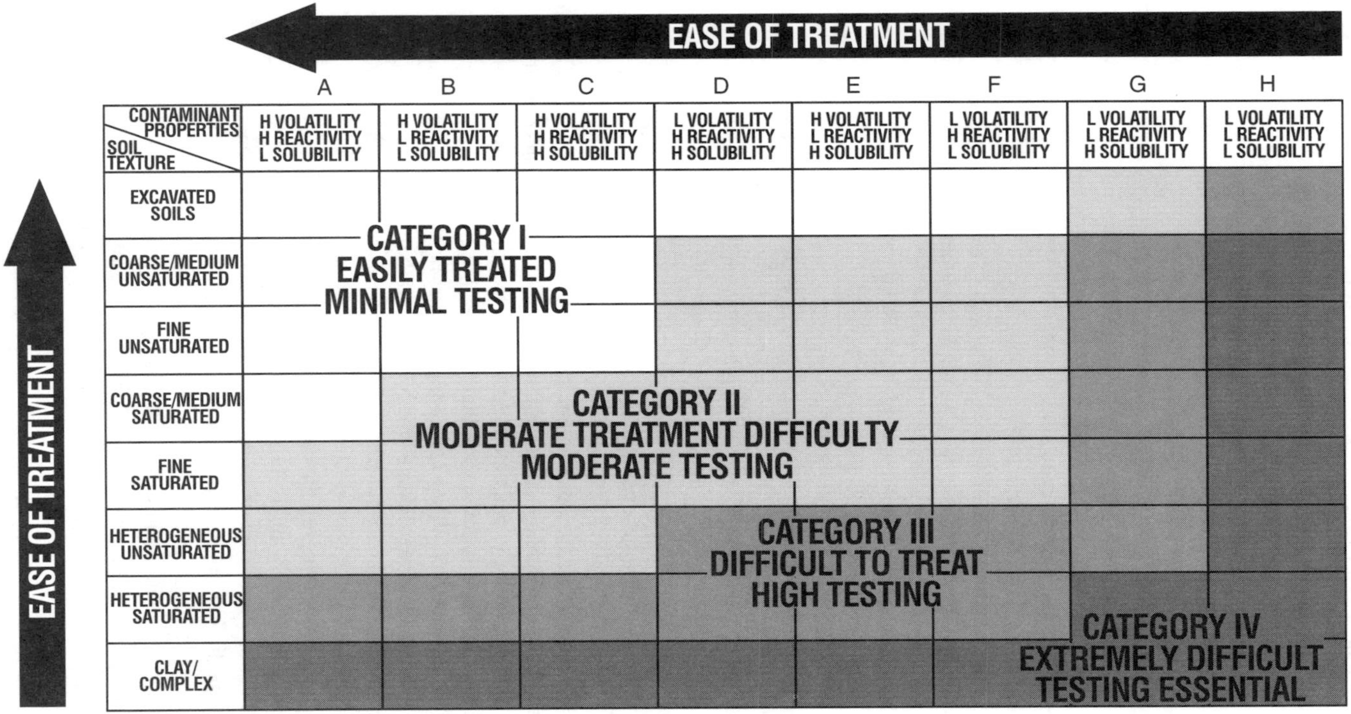

FIGURE 5-3 Treatability of contaminated sites and level of site-specific testing of remediation technologies required as a function of contaminant and geologic properties. Note that "H" indicates high and "L" indicates low volatility, reactivity, or solubility. (See Table 5-4 for a listing of sample contaminant compounds in each category.)

TABLE 5-4   Classes of Compounds Shown in Figure 5-3

Contaminant Class (as shown in Figure 5-3)	Volatility, Reactivity, and Solubility	Example Contaminants
A	HHL	Hydrocarbon fuels; benzene, toluene, ethylbenzene, and xylene
B	HLL	Trichloroethane, trichloroethylene, tetrachloroethylene
C	HHH	Acetone
D	LHH	Phenols, glycols
E	HLH	Methyl tertiary-butyl ether, tertiary butyl alcohol, methylene chloride
F	LHL	Naphthalene, small polycyclic aromatic hydrocarbons (PAHs), phthalates
G	LLH	Inorganic mixtures, metals of different chemistries
H	LLL	Polychlorinated biphenyls, pesticides, large PAHs

NOTES:
Volatility: High (H) > approximately 10 mm Hg; Low (L) < approximately 1 mm Hg
Reactivity: High - biodegradable, oxidizable; Low - recalcitrant
Solubility: High > approximately 10,000 mg/liter; Low < approximately 1,000 mg/liter

stage of testing does not assure success at a subsequent stage, and scaleup of the technology may be difficult.

While contaminant and hydrogeologic properties exert the primary influence on the amount of site-specific testing required prior to application of an innovative remediation technology, characteristics of the technology also influence the amount and detail of site-specific data that will need to be collected prior to installation of the technology. When technologies must be brought to the contaminant, more detailed site-specific data will be required than when the contaminant can be brought to the technology. In the first case, taking the technology to the contaminant, the subsurface properties must be detailed on a much finer scale than for the latter case, bringing the contaminant to the technology. Also, the site will need to be monitored much more intensively to prove that remediation is occurring. An example of this situation is use of a reactive treatment technology, such as bioremediation, for which the technology (bioremediation) is brought to the contaminants. To determine whether bioremediation is successful for a complex distribution of contaminants requires intensive monitoring and analysis. The presence of indigenous microbes and their biodegradation potential, the bioavailability of compounds, and the distribution of nutrients and moisture must be understood. Moreover, substances such as nutrients and electron acceptors must be delivered to the zones of contamination to support remediation, which might re-

TABLE 5-5 Site-Specific Testing Needs for Remediation Technologies

Category of Site	Data Needed to Determine Efficacy	Data Needed to Determine Applicability	Data Transferability	Commercialization Basis	Ease of Scaleup	Focus of Testing
Highly treatable (category I)	Fundamental properties of technology and site	Fundamental properties of technology and site	High for C-C[a], G-G, SU (for other sites in highly treatable category)	Fundamental properties and case histories	High	Testing for design
Moderately difficult to treat (category II)	Fundamental properties of technology and site	Field test data	Good for C-C, SU (for other sites in moderately treatable category)	Verified data and case histories	Moderate	Testing for application and design
Difficult to treat (category III)	Laboratory, bench, and pilot test data	Laboratory, bench, and pilot test data	Poor; requires site-specific testing	Verified data and site-specific testing	Moderate	Testing for efficacy, application, process verification, and design
Extremely difficult to treat (category IV)	Laboratory, bench, and pilot test data	Laboratory, bench, and pilot test data (several pilot tests)	Poor; requires site-specific testing	Site-specific testing	Poor	Testing for efficacy, application, process verification, and design

[a]C-C denotes contaminant-to-contaminant transfer of the technology; G-G denotes transfer of the technology from one type of site geology to another (within the same general category of treatability); SU denotes ease of data transfer in scaleup.

quire manipulating the flow field in a specific way to reach the contaminants. Determining whether any other processes, such as volatilization or mixing with natural waters, is acting to reduce concentrations will also be necessary. On the other hand, when contaminants can be brought to the technology, such as with reactive barriers, the amount of information needed and the application and analysis of the technology are, in relative terms, easier. In the case of reactive barriers, the flow field will need to be manipulated to deliver the contaminants to the barrier, but the manipulation will be on a much larger scale, which is inherently easier to do. Bringing the contaminants to the barrier generally requires less detailed site investigation because the focus is on flow field manipulations rather than on small-scale processes. In addition, technology assessment is much easier, requiring only a comparison of the concentrations of contaminants entering the barrier with the concentrations exiting the barrier, because the processes occurring in the barrier are known. Thus, approaches that bring the contaminants to the technology have an advantage in both the amount and type of data needed for site-to-site transfer and in the amount and type of data needed for evaluating the technology.

TECHNOLOGY PERFORMANCE VERIFICATION

The wide variation in methods used to assess the performance of innovative remediation technologies has made it very difficult for potential clients to judge the validity of remediation technology performance data. The "not tested in my backyard syndrome," in which owners of contaminated sites and regulatory personnel are reluctant to accept technology performance data from another site, is a significant problem in the remediation industry. In part, this reluctance stems from clients' concerns about potential regulatory or legal challenges to the selected remedy. Clients may fear that if they choose an innovative technology and their cleanup remedy is later legally challenged, proving in a court of law that the innovative remedy was, in fact, an adequate selection may be difficult. In deciding whether to admit scientific data into legal proceedings, courts of law must consider factors such as whether standards exist for the collection of such data, whether the data have received widespread acceptance in the scientific community, whether the data have been peer reviewed, and the potential for error in the data.[1] Meeting such legal standards may be difficult when innovative remediation technologies are chosen. Thus, a remediation technology developer may invest a great deal in a single field test hoping it will lead to additional customers, but a successful test often fails to lead to client acceptance of the technology, in part because of legal concerns.

---

[1]A recent Supreme Court case involving a toxic tort claim that Benedectin caused birth defects outlines the newest standards for the admissibility of scientific evidence in courts of law. See *Daubert v. Merrell Dow Pharmaceuticals, Inc.*, 113 S.Ct. 2786 (1993).

The fact that lack of credible performance data limits selection of innovative remediation technologies is now well recognized. Several efforts to develop protocols to standardize the testing, data collection, and regulatory approval process for remediation technologies are under way. Box 5-12 summarizes current programs in three categories: those for standardizing data reporting procedures, those for creating a more uniform regulatory approval process, and those for verifying technology performance. In the first category is the Federal Remediation Technologies Roundtable guide that federal agencies are to use in documenting cost and performance of remediation technologies. In the second category are programs by the western states, southern states, a six-state consortium, and Massachusetts to increase the level of regulator confidence in data on innovative remediation technology performance. In the third category are the SITE program (the oldest program for remediation technology verification) and the California Environmental Protection Agency Technology Certification Program.

Although the programs listed in Box 5-12 offer opportunities to report remediation technology performance data, independently verify these data, and specify steps necessary for regulatory approval of innovative remediation technologies, the existence of such a wide variety of programs in itself creates confusion for remediation technology developers and purchasers. Limited efforts to standardize the format for reporting cost and performance data under these various programs are under way, but nevertheless the different programs have different procedures for participation. Thus, the existence of these programs can exacerbate the problems faced by technology vendors in deciding which types of performance data to collect. Furthermore, the programs are voluntary and are not always accepted by agencies other than the ones participating in the program. Having a technology included in one of these programs may not provide a sales advantage except in the limited universe of sites under the jurisdiction of the agencies involved in the program. The costs of collecting all the data necessary for participation can be high, and technology developers may have to disclose company "secrets" in the process. Without the promise of a large market to make up for these costs, it is likely that very few companies will participate in the programs, except perhaps California's, which has a relatively large, well-defined market.

A uniform, widely used national program for testing and verifying the performance of new subsurface cleanup technologies is needed to provide a clear path for technology vendors to follow in planning how to prepare their technologies for the marketplace. The program should focus on verification of technology performance, meaning proving performance under specific conditions and providing assurance of data quality, rather than on certification, meaning guaranteeing technology performance. Because of the wide variation in contaminated sites, no technology can be guaranteed to achieve a given performance level at every site, and some degree of site-specific testing will always be required. However, having a uniform national protocol for reporting performance data and a mecha-

BOX 5-12
Testing, Verification, and Regulatory Approval Programs
for Remediation Technologies

Data Collection and Reporting Protocols

• *Federal Remediation Technologies Roundtable Guide to Documenting Cost and Performance for Remediation Projects:* The Federal Remediation Technologies Roundtable, a consortium of federal agencies involved in cleaning up hazardous waste sites, in 1995 published a guide specifying standard formats for documenting the performance of site cleanup technologies (Federal Remediation Technologies Roundtable, 1995). Agencies are required to use the guide to prepare cost and performance reports for Superfund sites on federal lands (Luftig, 1996). For information, contact the EPA's Technology Innovation Office, (703) 308-9910.

Regulatory Approval Protocols

• *Interstate Technology and Regulatory Cooperation (ITRC) Working Group:* The ITRC, a group initiated by the Western Governors Association, is developing regulatory approval protocols for several classes of hazardous waste remediation technologies. Most of the 27 states participating in the ITRC work group have agreed to accept remediation technology test results from other states if the tests are conducted according to the protocols the ITRC is developing. For information, contact the Western Governors Association, (303) 623-9378, or the ITRC's World Wide Web site, http://www.gnet.org/gnet/gov/interstate/itrcindex.htm.
• *Southern States Energy Board Interstate Regulatory Cooperation Project for Environmental Technologies:* The Southern States Energy Board is working to develop compatible regulations for environmental technologies in southern states. The project began with a pilot demonstration of data management and integration technologies in South Carolina and Georgia. For information, contact the Southern States Energy Board, (770) 242-7712.
• *Six-State Partnership for Environmental Technology:* Six states (California, Illinois, Massachusetts, New Jersey, New York, and Pennsylvania) in 1995 signed a memorandum to develop a process for the reciprocal evaluation, acceptance, and approval of environmental technologies. The partnership has begun this effort with pilot projects to review 12 different environmental technologies, including several for use in contaminated site remediation. For information, contact the New Jersey Office of Innovative Technology and Market Development at (609) 984-5418.

- *Massachusetts Strategic Envirotechnology Partnership (STEP):* STEP is a recently initiated program to promote use of new environmental and energy-efficient technologies in Massachusetts. Under the program, the state provides opportunities to pilot test technologies on state properties or at state facilities. The STEP program also helps expedite regulatory review and permitting of new environmental technologies using a team of innovative technology coordinators. In addition, it provides all technology developers in the program with a business plan review, including assistance in identifying potential markets and sources of funding. For information, call the Massachusetts Office of Business Development, (617) 727-3206.

Technology Performance Verification Protocols

- *Superfund Innovative Technology Evaluation (SITE) Program:* The first program for testing the performance of ground water and soil cleanup technologies, SITE was established in 1986 in response to a congressional mandate in the Superfund Reauthorization Act and Amendments (SARA) of 1986. SARA called for an "alternative or innovative treatment technology research and demonstration program." SITE is run by the EPA's National Risk Management Research Laboratory in Cincinnati. Under the program, EPA funds a select number of technology demonstrations each year. Technology developers can apply to have their technology tested under the SITE program by responding to an annual request for proposals. Developers pay for technology installation and operation costs; EPA pays for data collection and analysis. The SITE program, which has been criticized for failing to provide a market advantage to technologies that pass through it, received no funding in 1996, but funding was reinstated at $6 million in 1997. For information, contact the SITE program, (513) 569-7697.
- *California Environmental Protection Agency (CalEPA) Technology Certification Program:* In 1994, CalEPA established an environmental technology certification program in response to a mandate from the state legislature, specified in Assembly Bill 2060. The program will eventually provide mechanisms to certify all types of environmental technologies used in the state. The state's goal is to streamline the regulatory acceptance process for new environmental technologies and to increase customer confidence in performance data. The program began with a series of pilot tests to certify performance of a range of pollution prevention and environmental monitoring technologies. For information, contact CalEPA's Department of Toxic Substances Control, Office of Pollution Prevention and Technology Development, (916) 324-3823.

nism for reviewing the validity of the data would increase client and regulatory acceptance of credible performance data and would enable credible defense of the choice of an innovative remedy in courts of law. It would also facilitate the extrapolation of data from one site to another. The SITE Program, the only national program available for verifying remediation technology performance, has inadequate breadth, funding, and recognition to provide the needed level of remediation technology performance validation.

Verification of remediation technology performance should require reporting of data in the two categories described earlier in this chapter: (1) data showing that the technology works in reducing risks posed by specific contaminants under specific site conditions and (2) data linking the observed risk reduction with the technology. At least two types of evidence should be provided for each of these categories. The application for verification should provide a data summary sheet similar to the reports shown in Boxes 5-2, 5-3, 5-4, and 5-5. It should also specify the range of contaminant types and hydrogeologic conditions for which the technology is appropriate, and separate performance data should be provided for each different type of condition. Performance data should be entered in the coordinated remediation technologies data bases recommended in Chapter 3.

Three possible types of organizations could serve as the center of the verification program:

• *EPA:* The EPA SITE Program could be greatly expanded to allow for verification of a wide range of remediation technologies. Verification could be provided by EPA staff or contractors at EPA laboratories.

• *Third-party franchise:* A third-party center (under the direction of a private testing organization or professional association) could work with technology developers to establish test plans and conduct tests in the field or at a test facility, as appropriate. Staff of the center would evaluate the results and submit a verification report to the EPA.

• *Nonprofit research institute:* A nonprofit research institute affiliated with a university could establish technology evaluation protocols, either independently or based on guidelines from the EPA and other agencies. It could franchise other laboratories to assist with the testing and to evaluate results. These laboratories would then submit results to the institute for verification.

Regardless of which type of entity is responsible for verification, establishing a credible, widely used testing process will be essential. Questions regarding data acquisition, quality assurance and control, and appropriate measures of success would all need to be addressed. Whether data provided by the technology developer would be allowed in the verification process, or whether the data would need to be generated by an independent organization, would need to be established. The relative value of retrospectively and prospectively acquired data would need to be established. Roles of stakeholders (see Chapter 4) in the verification process would need to be defined. Incentives would need to be developed to participate in and use data produced by the program.

The verification program should be launched with a series of small pilot projects involving a variety of technology types, environmental media, and technology developers. The pilot programs would assist in checking whether the test protocols are adequate and in determining quality assurance and control procedures. In the pilot programs, technology vendors would draft a technology test plan in conjunction with the verification entity, which would either test the technology directly or oversee tests conducted by others. Verification of the results (or a decision not to verify the results) by the verification entity would follow.

DATA SHARING THROUGH
GOVERNMENT AND INDUSTRY PARTNERSHIPS

Private industries and government agencies "own" similar subsurface contamination problems. Yet, as discussed in this chapter, companies and agencies can be reluctant to accept remediation technology performance data generated by another company or agency. In addition to encouraging data acceptance through a verification program, sharing of data could be encouraged by forming technology testing and development partnerships including government agencies and a number of private companies. Such partnerships would, in the long run, provide cost savings to participating companies and agencies because they would leverage technology testing costs across a group of organizations so that no one organization would bear the entire cost.

One such partnership, the Remediation Technologies Development Forum (RTDF) already exists. The RTDF is an EPA-facilitated umbrella organization established in 1992. Through the RTDF, government and industry problem "owners" meet periodically to share information about problems of mutual concern and work together to find solutions (EPA, 1996b). The RTDF is currently supporting $20 million in work effort. Several formal RTDF teams are in place to develop innovative remediation technologies, and the RTDF is considering establishing more such teams (Kratch, 1997).

The first RTDF team formed is known as the Lasagna Consortium™. Through this partnership, Monsanto, DuPont, General Electric, the EPA, and the Department of Energy (DOE) are cooperating to develop a process that uses electroosmosis (see Box 3-3 in Chapter 3) to move contaminated ground water from low-permeability formations to in situ treatment zones. The EPA is supplying research capabilities, and the DOE is supplying funding and a test site at its Paducah, Kentucky, facility. The industrial partners are supplying program management, basic laboratory development, and design and construction capabilities. A successful pilot test to prove the principles underlying the technology's performance was completed in 1996, and a much-refined scaleup using zero-valent iron reaction zones (see Chapter 3 and Box 5-6) to destroy trichloroethylene is in progress.

A second RTDF team is investigating bioremediation of chlorinated solvents. The team consists of a consortium of six companies (DuPont, General Electric, Monsanto, Dow Chemical, ICI Zeneca, and Beak Consultants) working in partnership with the EPA, DOE, and Air Force. The consortium is investigating three different types of bioremediation: accelerated dechlorination, cometabolic bioventing, and intrinsic bioremediation. The DOE and Chlorine Chemical Council are providing funding, and the Air Force is providing test sites at the Dover, Delaware, Air Force Base. EPA is providing research in bioventing. The industrial partners are providing program management, laboratory studies, and design of the accelerated and intrinsic bioremediation protocols. Two pilot tests are under way, and work is being completed to select additional sites for a parallel series of pilot tests.

Recently established RTDF teams are demonstrating passive-reactive barriers for treatment of chlorinated solvents, in situ technologies for treating metals, and in situ techniques for cleaning up contaminated sediment. The RTDF is also establishing additional teams to investigate surfactant flushing systems for the treatment of DNAPLs in ground water and phytoremediation for the treatment of organic contaminants in soils.

The major driver behind the RTDF consortia is the desire to develop sound technologies that will reduce remediation costs to government and industry users. The close collaboration of those involved is leading to a shared understanding of the technologies. Participants hope that the effort will lead to early acceptance and application of the technologies, because three of the major stakeholders (technology users, developers, and regulators) are a party to the process. The EPA's participation has helped remove regulatory barriers to pilot testing.

It is too early to determine whether the RTDF arrangement will lead to rapid commercialization of the technologies being tested under the program. However, many elements are in place to speed the technologies through the pilot testing phase. For example, if the lasagna process proves successful, it is scheduled for full-scale implementation at Paducah, meaning there is a guaranteed first client for the technology. While such industry and government partnerships may not solve all the problems associated with testing and commercialization of remediation technologies, they should be encouraged as a potentially effective means for involving major stakeholders in mustering national resources to find solutions.

CONCLUSIONS

The wide variation in protocols used to assess the performance of innovative technologies for ground water and soil remediation has interfered with comparisons of different technologies and evaluation of performance data. In part because of the lack of standard performance reporting procedures, owners of contaminated sites and environmental regulators may hesitate to consider data from other sites in assessing whether an innovative remediation technology may be appropriate for their site. While a technology developer may invest large sums in conducting a field test to prove technology performance, potential clients may be hesitant to accept data from the field test if it was not carried out on the client's site and under the client's supervision.

The problem of variability in remediation technology performance data is now well recognized by environmental regulators, and various federal and state agencies have made efforts to standardize data collection and reporting procedures. However, the efforts of these agencies have not been coordinated. They thus provide little assurance to technology developers that following the procedures will provide a net benefit to the developer. The developer may expend large sums on testing a technology according to one agency's procedures, only to learn that the procedures will not be accepted by another agency. Some degree of national standardization in processes used to evaluate the performance of innovative remediation technologies is needed to allow for greater sharing of information, so that experiences gained in remediation at one site can be applied at other sites. In addition, more opportunities need to be created for cooperative technology development partnerships including government, industry, academia, and other interested stakeholders to encourage sharing and acceptance of data.

RECOMMENDATIONS

To standardize performance testing protocols and improve the transferability of performance data for innovative remediation technologies, the committee recommends the following:

•   In proving performance of an innovative remediation technology, technology developers should provide data from field tests to answer the following two questions:

1.   Does the technology reduce risks posed by the soil or ground water contamination?
2.   How does the technology work in reducing these risks? That is, what is the evidence proving that the technology was the cause of the observed risk reduction?

To answer the first question, the developer should provide two or more types of data leading to the conclusion that contaminant mass and concentration, or contaminant toxicity, or contaminant mobility decrease following application of the technology. To answer the second question, the developer should provide two or more types of evidence showing that the physical, chemical, or biological characteristics of the contaminated site change in ways that are consistent with the processes initiated by the technology, using evaluation procedures such as those shown in Table 5-1.

• **In deciding how much site-specific testing to require before approving an innovative remediation technology, clients and environmental regulators should divide sites into the four categories shown in Figure 5-3: (I) highly treatable, (II) moderately difficult to treat, (III) difficult to treat, and (IV) extremely difficult to treat.** For category I sites, site-specific testing of innovative remediation technologies should be required only to develop design specifications; efficacy can be determined without testing based on a review of fundamental principles of the remediation process, properties of the contaminant and site, and prior experience with the technology. For category II sites, field pilot testing should be required to identify conditions that may limit the applicability of the technology to the site; testing requirements can be decreased as the data base of prior applications of the technology increases. For category III sites, laboratory and pilot tests will be necessary to prove efficacy and applicability of the technology at the specific site. For category IV sites, laboratory tests and pilot tests will be needed, and multiple pilot tests may be necessary to prove that the technology can perform under the full range of site conditions.

• **All tests of innovative remediation technology performance should include one or more experimental controls.** Controls such as those summarized in Table 5-2 are essential for establishing that observed changes in the zone targeted for remediation are due to the implemented technology. Failure to include appropriate controls in the remediation technology performance testing protocol can lead to failure of the test to prove performance.

• **The EPA should establish a coordinated national program for testing and verifying the performance of new remediation technologies.** The program should be administered by the EPA and implemented by either EPA laboratories, a private testing organization, a professional association, or a nonprofit research institute. It should receive adequate funding to include the full range of ground water and soil remediation technologies and to test a wide variety of technologies each year. A successful test under the program should result in a guaranteed contract to use the technology at a federally owned contaminated site if the technology is cost competitive. The program should be coordinated with state agencies so that a technology verified under the program does not require additional state approvals.

• **Applications for remediation technology verification under the new verification program should include a summary sheet in standard format.** The summary sheet should contain information similar to that presented in Boxes 5-2, 5-3, 5-4, and 5-5. It should include a description of the site at which the technology was tested, the evaluation methods used to prove technology performance, and the results of these tests. It should also include a table showing the types of data used to answer each of the two questions needed to prove technology performance.

- **Applications for remediation technology verification should specify the range of contaminant types and hydrogeologic conditions for which the technology is appropriate.** Separate performance data should be provided for each different major class of contaminant and hydrogeologic setting for which performance verification is being sought.
- **Data gathered from technology performance tests under the verification program should be entered in the coordinated national remediation technologies data bases recommended in Chapter 3.** Data should be included for technologies that were successfully verified and for those that failed the verification process.
- **Technology development partnerships involving government, industry, academia, and other interested stakeholders should be encouraged.** Such partnerships can leverage resources to speed innovative technologies through the pilot testing phase to commercial application.

REFERENCES

Annable, M. D., P. S. C. Rao, W. D. Graham, K. Hatfield, and A. L. Wood. In press. Use of partitioning tracers for measuring residual NAPLs: Results from a field-scale test. Journal of Environmental Engineering.

Barker, J. F., G. C. Patrick, and D. Major. 1987. Natural attenuation of aromatic hydrocarbons in a shallow sand aquifer. Ground Water Monitoring Review (Winter):64-71.

Benson, R. C., and J. Scaife. 1987. Assessment of flow in fractured rock and karst environments. Pp. 237-245 in Karst Hydrogeology: Engineering and Environmental Applications, B. F. Beck and W. L. Wilson, eds. Boston: A. A. Balkema.

Cochran, W. G., and G. M. Cox. 1957. Experimental Design. New York: John Wiley & Sons.

EPA (Environmental Protection Agency). 1989. Evaluation of Ground-Water Extraction Remedies, Volume 2: Case Studies. EPA/540/2-89/054b. Washington, D.C.: EPA.

EPA. 1990. International Waste Technologies/Geo-Con In Situ Stabilization/Solidification, Applications Analysis Report. EPA/540/A5-89/004. Cincinnati, Ohio: EPA, National Risk Management Research Laboratory.

EPA. 1992. Evaluation of Ground-Water Extraction Remedies: Phase II, Volume I-Summary Report. Publication 9355.4-05. Washington, D.C.: EPA, Office of Solid Waste and Emergency Response.

EPA. 1996a. Accessing the Federal Government: Site Remediation Technology Programs and Initiatives. EPA/542/B-95/006. Washington, D.C.: EPA.

EPA. 1996b. Remediation Technologies Development Forum. 542-F-96-010. Washington, D.C.: EPA.

EPA. 1996c. State Policies Concerning the Use of Surfactants for In Situ Ground Water Remediation. EPA-542-R-96-001. Washington, D.C.: EPA.

ETI (EnviroMetal Technologies, Inc.). 1995. Performance History of the EnviroMetal Process. Internal Document. Guelph, Ont.: ETI.

Federal Remediation Technologies Roundtable. 1995. Guide to Documenting Cost and Performance for Remediation Projects. EPA-542-B-95-002. Washington, D.C.: EPA.

Feenstra, S., and J. A. Cherry. 1996. Diagnosis and assessment of DNAPL sites. In DNAPLs in Ground Water: History, Behavior, and Remediation, J. H. Pankow and J. A Cherry, eds. Portland, Ore.: Waterloo Press.

Gillham, R. W. 1995. Resurgence in research concerning organic transformations enhanced by zero-valent metals and potential application in remediation of contaminated groundwater. In Pre-prints of Papers Presented at the 209th American Chemical Society National Meeting, Anaheim, Calif., April 2-7, 1995. Washington, D.C.: American Chemical Society, Division of Environmental Chemistry.

Gillham, R. W., and S. F. O'Hannesin. 1994. Enhanced degradation of halogenated aliphatics by zero-valent iron. Ground Water 32:958-967.

Gillham, R. W., D. W. Blowes, C. J. Ptacek, and S. F. O'Hannesin. 1994. Use of zero-valent metals in in-situ remediation of contaminated groundwater. Pp. 913-930 in In-situ Remediation: Scientific Basis for Current and Future Technologies, W. G. Gee and N. R. Wing, eds. Columbus: Battelle Press.

Hinchee, R. E., and M. Arthur. 1991. Bench-scale studies of the soil aeration process for bioremediation of petroleum hydrocarbons. Journal of Applied Biochemistry and Biotechnology 28/29:901-906.

Hinchee, R. E., S. K. Ong, R. N. Miller, D. C. Downey, and R. Frandt. 1992. Test plan and technical protocol for a field treatability test for bioventing. Brooks Air Force Base, Texas: U.S. Air Force Center for Environmental Excellence.

Jin, M., M. Delshad, V. Dwarakanath, D. C. McKinney, G. A. Pope, K. Seperhnoori, C. Tilburg, and R. E. Jackson. 1995. Partitioning tracer test for detection, estimation, and remediation performance assessment of subsurface nonaqueous phase liquids. Water Resources Research 31(5): 1201-1211.

Johnson, P. C., R. L., Johnson, D. Neaville, E. Hansen, S. M. Stearns, and I. J. Dortch. 1997. Do conventional monitoring practices indicate in situ air sparging performance? Pp. 1-20 in In Situ and On-Site Bioreclamation, 3rd Symposium. Washington, D.C.: American Chemical Society.

Kratch, K. 1997. Consortium researches technologies to solve priority problems. Water Environment and Technology 9(2):32.

Lahvis, M. A., and A. L. Baehr. 1996. Estimation of rates of aerobic hydrocarbon biodegradation by simulation of gas transport in the unsaturated zone. Water Resources Research 32(7):2231-2249.

Laws, E. P. 1996. Letter to Superfund, RCRA, UST, and CEPP national policy managers, Federal Facilities Leadership Council, and brownfields coordinators on EPA initiatives to promote innovative technology in waste management programs. April 29. Washington, D.C: EPA.

Luftig, S. D. 1996. Documenting Cost and Performance Data for Site Remediation Projects: January 17 Memorandum. Washington, D.C.: EPA, Office of Solid Waste and Emergency Response.

National Research Council. 1993. In Situ Bioremediation: When Does It Work? Washington, D.C.: National Academy Press.

National Research Council. 1994. Alternatives for Ground Water Cleanup. Washington, D.C.: National Academy Press.

Orth, W. S., and R. W. Gillham. 1996. Dechlorination of trichloroethene in aqueous solution using Fe^0. Environmental Science & Technology 30:66-71.

Rao, P. S. C., M. D. Annable, R. K. Sillan, D. P. Dai, K. Hatfield, W. D. Graham, A. L. Wood, and C. G. Enfield. In review. Field-scale evaluation of in-situ cosolvent flushing for remediation of an unconfined aquifer contaminated with a complex NAPL. Water Resources Research.

Roberts, P.V., G. D. Hopkins, D. M. Mackay, and L. Semprini. 1990. A field evaluation of in situ biodegradation of chlorinated ethenes: Part 1—Methodology and field site characterization. Groundwater 28:591-604.

Semprini, L., P. V. Roberts, G. D. Hopkins, P. L. McCarty. 1990. A field evaluation of in situ biodegradation of chlorinated ethenes: Part 2—Results of biostimulation and biotransformation experiments. Groundwater 28:715-727.

Siegrist, R. L., O. R. West, M. I. Morris, D. A. Pickering, D. W. Greene, C. A. Muhr, D. D. Davenport, and J. S. Gierke. 1995. In situ mixed region vapor stripping in low-permeability media. 2. Full scale field experiments. Environmental Science & Technology 29:2198-2207.

Starr, R. C., J. A. Cherry, and E. S. Vales. 1993. Sealable joint sheet pile putoff walls for preventing and remediating groundwater contamination. In Proceedings of the Technology Transfer Conference. Toronto: Ontario Ministry of the Environment.

Steinberg, D. M., and W. G. Hunter. 1984. Experimental design: Review and comment. Technometrics 26(2):71-97.

Thomas, A. O., D. M. Drury, G. Norris, S. F. O'Hannesin, and J. L. Vogan. 1995. The in situ treatment of trichloroethene-contaminated groundwater using a reactive barrier—results of laboratory feasibility studies and preliminary design considerations. In Contaminated Soil '95, W. J. van den Brink, R. Bosman, and F. Arendt, eds. Amsterdam: Kluwer Academic Publishers.

West, O. R., R. L. Siegrist, J. S. Gierke, S. W. Schmunk, A. J. Lucero, and H. L. Jennings. 1995. In situ mixed region stripping in low-permeability media. 1. Process features and laboratory experiments. Environmental Science & Technology 29:2191-2197.

6

Comparing Costs of Remediation Technologies

As with any product marketed in a competitive environment, information about the costs of innovative remediation technologies is as important in determining their ultimate commercial success as are performance data. The potential client wishes to choose the most cost-effective technology and, before selecting an innovative technology, will require some method to measure its cost against other available options on some standardized basis. Technology developers and investors need to have reliable cost information to determine whether the technology will be profitable.

Because of differences in site conditions, establishing cost data for innovative remediation technologies can be difficult, especially for in situ processes. Even if capital and operating costs have been established for candidate remediation technologies, the way in which these costs were developed and the way in which they are expressed may lead to quite different conclusions about the relative economic merits of the technologies. The client's view of the relative cost of remediation options, in turn, has implications for the remediation technology provider and ultimately determines which technologies will move forward from development to commercial success.

This chapter recommends a strategy for developing and analyzing cost data to allow valid comparisons of different types of remediation technologies.

LIMITATIONS OF EXISTING COST REPORTING STRUCTURES

For a variety of reasons, it is currently difficult to impossible to develop accurate comparisons of remediation technology costs in many situations.

One of the most significant problems with developing cost information is that costs reported under a set of conditions at one site are very difficult to extrapolate to other sites. Like technology performance, technology costs are sensitive to site-specific geologic, geochemical, and contaminant conditions, especially for in situ technologies.

A second problem is that technology vendors may report costs using a variety of different metrics that cannot be compared directly. Costs may be reported as dollars per volume treated, reduction in contaminant concentrations achieved, contaminant mobility reduction achieved, mass of contaminant removed, or surface area treated. For example, costs of a physical wall for containing or treating contaminants in place may be reported as dollars per area of wall surface, while the costs for a pump-and-treat system may be reported as dollars per volume of ground water treated. Such variations in cost reporting metrics make it difficult to compare costs of competing technologies using data from previous applications at different sites.

A third problem is that often technology providers do not report the variable costs, such as permitting, mobilization of equipment to the contaminated site, treatability studies to prove the technology or obtain permits, and system design or modification for site conditions. Just the "up and running" costs are given. This may be acceptable if the user only wants to compare the cost of installed operations, but the user is usually interested in the overall project cost. If certain remediation technologies have large and variable initial costs, they may not be competitive, even if the "up and running" costs appear competitive.

A fourth problem is inconsistencies in the way costs are derived. Comparisons of unit costs have little meaning unless there is uniformity in the underlying methodologies and assumptions used in calculating the costs. For example, if different interest rates are used to estimate the costs of a cleanup system over its entire life cycle, the conclusions about the cost competitiveness of a technology can vary widely.

A final problem is that for in situ technologies, cost information is often developed by geotechnical consultants rather than technology providers and is rarely compiled for general reference by the private user. This loss of compiled cost information greatly hinders the dissemination of consistent cost information and makes it difficult for a new technology provider to develop comparative cost information. Further, even where cost information is made available to private users, it is extremely rare to see detailed cost breakdowns that would allow the reviewer to judge the realism of the cost elements. While the federal government is beginning to compile cost data and create guidelines for cost computation and reporting at federal sites (Federal Remediation Technologies Roundtable, 1995), these guidelines have not been adopted by the private sector.

OPTIMIZING THE DEVELOPMENT AND
REPORTING OF COST DATA

The inherent uncertainties associated with the subsurface environment present unique challenges to those who wish to compare the costs of remediation technologies. Costs of remediation technologies will never be comparable in the same way that the cost of devices whose performance is uniform in every circumstance can be compared. Nevertheless, a variety of steps can be taken to enable technology providers and users to gather information on the costs of different remediation options and develop meaningful cost comparisons to evaluate the options.

Development of Template Sites for Cost Comparisons

The most difficult problem in developing sound, comparable cost information is in the application of in situ remediation technologies, for which site-specific conditions determine the way in which a technology (or technique) is applied. While some metric is necessary to capture and compare the costs of in situ technologies, it will be sensitive to the site-specific hydrogeology and contaminant conditions of the site, and so some description of this situation should accompany the cost information. Unfortunately, even if this is done, it is still difficult to compare costs between sites.

One way to overcome the problems associated with comparing cost data from different sites is to develop a set of "template sites" that can be used to compare relative costs of different classes of technologies. Each template site would have standard dimensions and hydrogeologic properties, and the template could be adapted to estimate costs for different types of contaminants and remediation goals. Table 6-1 shows basic parameters for eight types of site templates that could be developed to represent a range of conditions of aquifer depth, thickness, and permeability and ground water flow rate. (Excluded from consideration are fractured rock aquifers, which are a special case requiring site-specific analysis.) In addition to the parameters shown in the table, assumptions need to be made about the dimensions of the contaminated area. These dimensions can be highly variable, but for estimating purposes, the plume can be assumed to be spreading from a source area 64 m (210 ft, or the side of 1-acre square surface area) transverse to the flow direction. These conditions can be used to estimate the end points of the cost range for a technology, rather than to specify one "typical" cost. Additional templates could be constructed to provide midpoints in the cost range, but this would increase the amount of work involved in using the templates. For each template, detailed information such as that shown in Table 6-2 would also need to be specified.

While the lists in Tables 6-1 and 6-2 are not all inclusive, they contain critical elements that influence the costs of ground water remediation technologies.

TABLE 6-1 General Parameters for Template Sites for Comparing Costs of Ground Water Remediation Technologies

Template Number	Depth to Water Table (m)	Aquifer Thickness (m)	Aquifer Permeability (cm/sec)	Ground Water Flow Rate (m/year)
1	4.6 (15 ft)	7.6 (25 ft)	5.0×10^{-4}	3 (10 ft/yr)
2	4.6 (15 ft)	7.6 (25 ft)	2.5×10^{-2}	150 (500 ft/yr)
3	4.6 (15 ft)	21 (70 ft)	5.0×10^{-4}	3 (10 ft/yr)
4	4.6 (15 ft)	21 (70 ft)	2.5×10^{-2}	150 (500 ft/yr)
5	30 (100 ft)	7.6 (25 ft)	5.0×10^{-4}	3 (10 ft/yr)
6	30 (100 ft)	7.6 (25 ft)	2.5×10^{-2}	150 (500 ft/yr)
7	30 (100 ft)	21 (70 ft)	5.0×10^{-4}	3 (10 ft/yr)
8	30 (100 ft)	21 (70 ft)	2.5×10^{-2}	150 (500 ft/yr)

NOTE: Soil porosity is assumed to be 25 percent, and hydraulic gradient is assumed to be 0.005 cm/cm for all eight cases.

TABLE 6-2 Detailed Information Needed for Template Sites for Comparing Costs of Ground Water Remediation Technologies

Site Characteristics
- Conditions of site access
- Access to power, utilities
- Vadose zone soil classification
- Soil classification of ground water-bearing zone to be remediated
- Dimensions of contaminated zone; volume of contaminated area

Contaminated Ground Water Characteristics
- pH, dissolved oxygen concentration
- Total dissolved solids concentration, hardness, iron concentration, manganese concentration, concentrations of other potential foulants
- Redox potential
- Soil adsorption/desorption properties

Contaminant Characteristics
- Contaminant concentration profile
- Character/quantity of source materials (DNAPL, etc.)

NOTE: This table assumes that the general aquifer characteristics shown in Table 6-1 have already been specified.

The elements shown in the tables will not cover all remediation techniques and possible site scenarios, such as the existence of major heterogeneities in the geologic formation. Such templates may overestimate performance and underestimate costs for heterogeneous sites, and the nature of the effects of heterogeneity on remediation technology performance and costs is poorly understood. However, for most technologies, such templates will provide a consistent basis for estimating order-of-magnitude upper- and lower-bound costs for alternative technologies. The Department of Energy and a few private companies use this general method when comparing technology alternatives against a baseline technology (Ellis, 1996; Herriksen and Booth, 1995).

An example of the potential use of a template site would be to generate first approximation costs for comparing different dense nonaqueous-phase liquid (DNAPL) treatment and removal technologies. Figure 6-1 shows a plot plan for a sample template site used to compare costs for cleanup of a DNAPL spill of 680 kg (1,500 lbs) of perchloroethylene (PCE) covering an area of 0.4 ha (1 acre). The template site has a confining layer 12 m (40 ft) below the surface and an

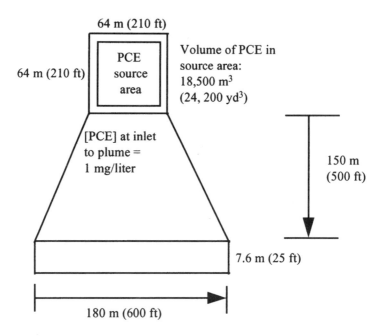

FIGURE 6-1   Example of a template site that could be used to compare the costs of remediation technology alternatives. In this example, 680 kg (1,500 lbs, or 111 gal) of nonaqueous-phase PCE have spilled over a 0.4-ha (1-acre) area. The depth to the water table is 4.6 m (15 ft), and the aquifer thickness is 7.6 m (25 ft). The porosity is 0.25; ground water velocity is 150 m/year (500 ft/yr). These parameters correspond to template 2 in Table 6-1.

unconfined surficial aquifer 4.6 m (15 ft) below the surface. The aquifer porosity is 0.25, and the ground water velocity is 150 m/year (500 ft/year). Based on hydrodynamic dispersion parameters typical for a homogeneous aquifer, a plume of dissolved contaminants would have spread laterally 180 m (600 ft) at a distance of 150 m (500 ft) downgradient from the source zone. The computations assume that the plume exits the DNAPL area at a concentration of 1 mg/liter (approximately 0.7 percent of the aqueous solubility). This would result in about 680 kg (1,500 lbs) of PCE being flushed through a 180-m-wide (600-ft-wide) "technology alternative" zone downgradient of the source zone, where the costs of plume treatment or containment options could be compared over a 30-year period. If the source zone and plume dimensions and characteristics are realistically defined, it would also be possible to compare source treatment options to plume remediation options, assuming both options achieve acceptable levels of environmental protection and that any differences in level of protection are specified in the cost comparisons. This approach allows comparison of remediation options, but it also helps in the evaluation of the importance of many of the variables that are used to design and select remediation technologies.

Within the source zone, if surfactant flushing were used, costs could be estimated by determining the surfactant injection and withdrawal system needs, the amount of surfactant needed to mobilize the DNAPL, the time needed for the surfactant to be recirculated, and all the costs associated with the operation (including treatment of pumped fluids). Similar estimates could be prepared for thermally enhanced soil vapor extraction or for simple containment using a perimeter slurry wall. At the downgradient point in the plume, alternatives such as a permeable reaction wall or pump-and-treat system could be evaluated. The cost of each technology would be estimated for the expected lifetime of the remediation, or up to 30 years, where the net present cost increase becomes very small.

While the site template's parameters are somewhat arbitrary, they can be developed to create a realistic and consistent basis for cost comparisons. The template can be modified to fit a specific class of contamination problems. Actual performance information gathered from field tests or prior full-scale applications of the technology can provide the appropriate inputs to the cost model. In designing template sites, caution must be used to ensure that the templates are sufficiently general to allow inclusion of a range of technologies with similar capabilities but specific enough to produce reasonable cost comparisons. It is also important to specify the environmental end point achieved by the remediation technology and to compare costs based on achieving equivalent end points.

Although development of template sites for cost comparisons adds another level of complexity to the analysis of technology performance, this strategy will help reduce the problems associated with comparing in situ technologies. Owners of contaminated sites (including corporations and government agencies) and technology developers should cooperate to establish a procedure, possibly implemented through a ground water remediation cost definition working group, to develop and refine a system of template sites for comparing costs of ground water remediation options. Similar templates should be developed for comparing costs of alternative soil remediation options. A working group might be convened under the auspices of the Remediation Technologies Development Forum or the American Academy of Environmental Engineers, which recently completed monographs with guidelines on how to apply eight innovative remediation technologies. Computer models of the templates could be created to make them relatively easy for technology developers to use. The models could be pilot tested and then modified and improved over time. Once the templates are developed, owners of contaminated sites and federal agencies should require that cost information from technology suppliers be presented to them in a template format if the technologies are to be evaluated for purchase.

Use of Standard Metrics for Cost Reporting

Unit remediation costs are the distillation of the complex process of cost development. They are the most basic way of expressing a technology cost using some common metric. Table 6-3 shows examples of common cost measures. Cost measures reported by technology providers vary depending on whether the technology treats the contaminants in situ or ex situ, whether it is designed for containment or remediation, and whether the contaminated material is soil, other solid material, or ground water. The different metrics used to report costs of different types of technologies make it very difficult, in some cases nearly impossible, to compare the costs of different contaminant management options.

TABLE 6-3   Common Unit Measures of Cost

Matrix	Ex Situ	In Situ
Soil/waste material	Cost/volume ($/m^3, $/yd^3)*	Cost/volume ($/m^3, $/yd^3)
	Cost/weight ($/tonne, $/ton)	Cost/weight of contaminant treated or removed ($/kg, $/lb)
	Cost/weight of contaminant treated or removed ($/kg, $/lb)	Cost/vertical wall area ($/m^2, $/ft^2)
		Cost/surface area ($/ha, $/acre)
Ground water	Cost/volume ($/m^3, $/1,000 gal)*	Cost/volume ($/m^3, $/1,000 gal)
	Cost/weight of contaminant treated or removed ($/kg, $/lb)	Cost/weight of contaminant treated or removed ($/kg, $/lb)
		Cost/surface area ($/ha, $/acre)

*Often used in reports prepared by the Superfund Innovative Technology Evaluation Program.

For ex situ technologies, technology providers typically report costs in terms of dollars per volume treated. Sometimes this information is supplemented by percent reduction in contaminant concentration or mobility achieved. For in situ technologies, cost reporting is less standard. Cost per volume treated is often used, but the precise volume treated may not be known. The cost of hydraulic containment systems is sometimes reported using the same measures of cost as are used for pump-and-treat systems, but costs of physical containment devices are reported using a quite different set of metrics. Capping costs are typically reported as area of surface capped, while side wall system costs are reported as cost per area of wall surface. Whether the technology is in situ or ex situ or a containment system, the cost per volume of material treated or contained is rarely reported.

While the type of cost information most helpful to technology users may vary with the type of technology and contaminated media, nevertheless technology providers should always provide certain basic information to allow comparisons of the costs of different types of technologies. As a general rule, unit cost metrics for both in situ and ex situ technologies should include both a cost per unit volume of the contaminated matrix treated and also a cost for the mass of the specific contaminants removed, treated, or contained. Information should also be supplied on the starting concentrations of contaminants of concern as well as the percent removal, destruction, or containment achieved. In many instances, unit costs vary with the size of the remediation project and, in the case of processing equipment, the size and throughput of the unit. Whenever unit remediation costs are presented, the technology provider should report the amount of material remediated as well as the process rate.

Documentation of Costs Using Consistent Procedures

Much of the uncertainty in evaluating the costs of remediation technologies, especially in comparing an emerging technology against an established one, is brought about by inconsistencies in the way the costs are derived and reported. Different assumptions used in calculating costs can lead to vastly different conclusions about the relative economic merits of one technology versus another, yet it is extremely rare to see detailed cost breakdowns that would allow the reviewer to judge the realism of the cost elements.

The problem of inconsistencies in cost derivation and reporting has two elements. First, cost estimators use different assumptions about what cost elements should be included in the total estimate and how much detail should be included in reporting these elements. Second, estimators make different assumptions about interest rates. Better standardization and documentation of cost development procedures could help solve both of these problems.

Cost Element Selection

At the most basic level, remediation technology developers need to use standard cost elements in computing the total cost of a technology, and they need to document these elements in a standard format. Cost computations should show one-time start-up costs, such as studies to prove the technology or obtain permits, as well as the capital and operating costs of the technology itself. Table 6-4 shows the types of cost elements that need to be included in capital and operating cost estimates (Herriksen and Booth, 1995).

The Federal Remediation Technologies Roundtable has developed a guide to documenting the costs of remediation projects carried out at federal facilities (Federal Remediation Technologies Roundtable, 1995). This guide might serve as a starting point for standardizing the reporting of cost elements, but it will need refinement. The guidelines are based on use of the federal "work breakdown structure" (WBS), a system set up to catalog the individual cost elements of a project in great detail.

The WBS has several levels of detail, with the highest level (level 1) having the least detail. For remediation, level 1 simply specifies "hazardous, toxic, and radioactive waste remedial action." Level 2 lists a number of specific remediation activities (such as "mobilization and preparatory work" and "monitoring, testing and analysis") and a series of specific remediation techniques (such as "stabilization/encapsulation"). Level 3 provides additional sub-elements of detail, including "mobilization of personnel" and "mobilization of construction equipment and facilities" under the level 2 "mobilization and preparatory work." For other government accounting purposes, level 4 contains very detailed information used to assemble a cost estimate, but for remediation, level 4 only distinguishes between portable and permanent treatment units. Level 5 is a compilation of general portable unit treatment cost elements (such as "solids preparation and handling") at a degree of detail similar to that of level 3.

The remediation WBS system seems more appropriate for compiling costs of procured services than for setting a framework for developing standardized costs of new or developing technologies. The WBS by its nature compiles costs into specific technology categories and uses a standard list of known, specific cost elements for existing practices. Consequently, it may not be appropriate for compiling costs for new technology developments.

No system can account in advance for every detailed cost element of a technology, but a general framework for developing costs should be used. Because most organizations outside of the government complex do not compile their costs according to the WBS (although they may consider similar cost elements), some flexibility will be needed in cost documentation. The cost elements in Table 6-4 should be used as the minimum basis in developing cost comparisons among technologies. Where appropriate, technology developers and consultants should

TABLE 6-4   Typical Cost Categories Used to Compile or Estimate Costs

Capital Costs	Operating Costs

Capital Costs

A. Site preparation*
   Site clearing
   Site access
   Borehole drilling
   Permits/licenses
   Fencing
   Heat, gas, electricity, and water to
      install system

B. Structures*
   Buildings
   Platforms
   Equipment structures
   Equipment shed/warehouse

C. Process equipment and appurtenances*
   Cost of technology parts and supplies
   Materials and supplies to make
      technology operative

D. Non-process equipment*
   Office and administrative equipment
   Data processing/computer equipment
   Safety equipment
   Vehicles

E. Utilities*
   Plumbing, heating, light, security, and
      vent equipment

F. Labor*
   Direct labor necessary to acquire,
      mobilize, and install system
   Supervisory and administrative labor to
      acquire, mobilize, and install system
   Design and engineering

G. Other
   Rental of commercial equipment to
      mobilize and install system
   Start-up and testing

Operating Costs

A. Direct labor*
   Direct labor to operate equipment
   Direct labor supervision
   Payroll expenses (FICA, vacation,
      worker medical insurance, pension
      contribution)
   Contract labor
   Maintenance direct labor

B. Direct materials
   *Consumable supplies**
   Process materials and chemicals
   Utilities
   Fuels
   Replacement parts

C. Overhead
   Plant and equipment maintenance
   Liability insurance
   Shipping charges
   Equipment rental for operations
   Vehicle supplies and insurance
   Transportation
   Licensing

D. General and administrative
   Administrative labor
   Marketing
   Communications
   Project management
   Travel expenses
   Interest expenses

E. Site Management
   Maintenance contract for equipment
   *Waste disposal**
   Health and safety requirements
   Contract services
   Site closure activities
   *Analytical services**
   *Demobilization**
   Regulatory reporting

*Information typically supplied in reports from the Superfund Innovative Technologies Evaluation Program.

SOURCE: Herriksen and Booth, 1995.

structure reporting according to level 3 of the WBS, taking into consideration the elements of levels 4 and 5 of the remediation WBS.

In order to facilitate the process of cost development, the ground water remediation cost definition process or working group (discussed above) should establish a clearly defined general framework of important cost elements. The framework must suit the needs of the private technology development community. The role of the WBS in standardizing remediation technology costs should be re-evaluated, and the WBS process should be documented in a form that facilitates better understanding and use by the private sector.

Present Worth Calculation

In documenting the capital and operating costs of a remediation technology, developers need to indicate clearly their assumptions about interest rates and taxes. Developers should also tailor these assumptions to the needs of the technology user, which will vary depending on whether the client is a private company or a government agency.

Interest rate assumptions affect computations of the total cost of a technology because of the time value of money. That is, because cash in hand can be reinvested, it is more valuable than an equal amount of cash to be generated in the future. Equivalently, costs that can be deferred have less impact on a company's bottom line than costs that must be paid immediately. Financial analysts compare cash flows in different time periods by discounting them to present value at some discount rate, which may be the cost of capital for a business or the cost of borrowing for a government entity, according to the following equation:

$$PC = \sum_{n=0}^{y} \frac{CE_n (1+i)^n}{(1+k)^n}$$

where

PC = present cost
CE_n = cash expenditure in year n in present dollars
k = cost of capital or discount rate
n = year in which costs are incurred
y = total number of years of expected expenditures
i = inflation rate

For example, at a discount rate of 12 percent, a $100 payment to be used two years hence is equivalent to a payment of $80 today, assuming zero inflation.

Businesses have many different ways to estimate their cost of capital. A company seeks to earn a return on its investments, which will cover interest expense on borrowings and provide an attractive total return (dividends plus stock appreciation) to shareholders. A company may also view its cost of capital in terms of the rate of return that might be achieved if the money were invested in an available project of high return. The cost of capital is an approximation of the return levels required to achieve these objectives and hence may be considered significantly greater than the cost of debt alone. Unlike businesses, government agencies consider only the cost of debt in computing the present cost of future expenditures. So, while a business might assume a discount rate of 12 percent or higher, a government agency would typically use a rate of only 6 percent.

Businesses are able to obtain tax credits against the cost of remediation equipment. The effects of income tax considerations often vary widely from one investment alternative to another, so it is generally imperative for a business to compare the relative economics of remediation alternatives on an after-tax basis to have a valid economic analysis. Most U.S. corporations choose to benchmark their performance on an after-tax operating income basis. They calculate this basis by taking the difference between sales revenue and operating costs, capital costs, and income taxes. Because remediation costs are generally considered operating expenses for a corporation, they are deducted from the revenues, lowering the amount of taxes that a company owes. For U.S. corporations, the federal and state combined effective corporate tax rate ranges from to 35 to 40 percent (Stermole and Stermole, 1996). Businesses refer to the present cost of an item on an after-tax basis as the "net present cost."

As shown in the examples in Boxes 6-1 and 6-2 and the accompanying figures, the different assumptions that government agencies and private companies make about discount rates and tax liability can lead to quite different determinations of the net present costs of different technologies. In the example in Box 6-1, the net present cost that a business would compute for a pump-and-treat system operating over 30 years is $1,684,000, while the present cost that a government agency would calculate is $4,060,000. In the example in Box 6-2, different assumptions about discount rates and taxes would lead a government agency to conclude that an accelerated bioremediation system requiring 5 years to complete a cleanup would be more cost-effective than using intrinsic bioremediation over 30 years, while a business would reach the opposite conclusion. Thus, financial performance measures are powerful tools in strategic technology development and planning, but they should not be used mechanically. In the example in Box 6-1, the prospective client would find the pump-and-treat case more attractive than might be assumed by a technology developer using different assumptions. Similarly, in the example in Box 6-2, a technology developer using government discount rate assumptions might misread the market by concluding that businesses would view intrinsic bioremediation technologies as expensive and not competitive. It is very important that the provider use realistic measures of the cost of the technology versus alternatives when deciding if the new technology will be competitive with others, as judged by the user.

BOX 6-1
Present Cost Calculations: Government Versus Business

To calculate the net present cost of a pump-and-treat system operating over an extended time period, a typical business might use an inflation rate of 3 percent and a discount rate of 12 percent. The business would also deduct from the costs the tax credit obtained by building the treatment system. A government agency (and some technology providers), on the other hand, would typically use a discount rate of 6 percent and would not consider taxes. If the pump-and-treat system has an initial capital cost of $1 million and an annual operating cost of $150,000, then the business and government agency calculations for the total cost during the first three years would differ as follows:

	Government Cost Basis ($ thousands)	Business Cost Basis ($ thousands)
Year 1		
Equipment cost	1,000	1,000
O&M cost	150	150
Total year 1 cost	1,150	1,150
Total year 1 cost after taxes	1,150	713 (38% tax)
Discounted net present cost	1,150	13
Year 2		
Equipment cost	0	0
O&M cost (3% inflation)	155	155
Total year 2 future cost	155	155
Total year 2 after taxes	155	96 (38% tax)
Discounted net present cost	146 (6% disc.)	86 (12% disc.)
Year 3		
Equipment cost	0	0
O&M cost (3% inflation)	159	159
Total year 3 future cost	159	159
Total year 3 after taxes	159	99
Discounted net present cost	142 (6% disc.)	79 (12% disc.)
Total cost for years 1-3 on present cost basis	1,437	877

In subsequent years, the calculations would follow in the same manner, and the total cost estimates would continue to diverge.

Figure 6-2 shows the cumulative present cost for these cases extended to 30 years. As shown in the figure, the cumulative present cost as computed by the government is more than twice that computed by the business. The cumulative cost that a business might calculate would level off rapidly, with 90 percent of the total cost incurred by year 18. On the other hand, the government cumulative cost calculation would continue to climb rapidly, reaching the 90 percent expenditure point in year 25.

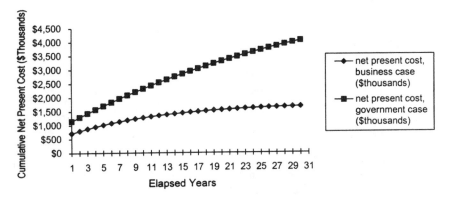

FIGURE 6-2   Business net present cost estimate versus government net present cost estimate for a hypothetical pump-and-treat system.

Inclusion of Cost Data in National Technology Performance Data Bases

Once cost information for a technology is developed, it should be made available to other potential technology users. The coordinated national data bases on remediation technologies recommended in Chapter 3 should include information on technology costs. This cost information should be reported in the data bases using the guidelines recommended above. The data bases should provide a description of template sites useful for cost comparisons. For each technology, the data bases should include separate cost data for each type of template site and contaminant type for which the technology is appropriate. The data bases should report costs as dollars per unit volume of the contaminated matrix treated and dollars per mass of the specific contaminant removed. It should include starting concentrations of the contaminants along with the dollar figures. Capital and operating costs should be reported separately, allowing users to prepare estimates of the present costs using discount rate assumptions appropriate for their circumstances.

BOX 6-2
How Financial Assumptions Affect Technology Selection

Differences in financial assumptions, as shown in Box 6-1, can lead to substantially different conclusions about which of two technologies is the most cost competitive. Figure 6-3 compares the present cost for a hypothetical accelerated bioremediation technology that would clean up a contaminated plume in 5 years to a hypothetical intrinsic bioremediation method in which slower natural degradation processes would also result in plume decontamination, but over a 30-year period. For the accelerated case, the initial equipment cost is $750,000, and the annual cost for nutrient addition is $200,000 per year over 5 years. For the intrinsic case, the initial equipment cost is $300,000 (for monitoring wells), and annual monitoring costs for making sure the contaminants are degrading are $100,000 per year over 30 years. The present costs in both situations use the typical business assumptions about discount rates and taxes presented in Box 6-1. Assuming both options result in the same environmental end point, in this example remediation would be less costly using the slower technology (although there may be other costs, such as delaying potential sale of the land, that would need to be considered). The market for the accelerated technology might not be as large as the developer would expect if the potential users conclude they could use intrinsic bioremediation in a significant number of situations.

If the same computations are made for the two cases using the 6 percent discount rate that a government-oriented developer might use with no consideration of taxes, the outcome would be considerably different:

	Intrinsic Bioremediation (30 years)	Accelerated Bioremediation (5 years)
"Government" estimate (3% inflation, 6% discount rate, no taxes)	$2,340,000	$1,695,000
"Business" estimate (3% inflation, 12% discount rate, 38% corporate tax rate)	$895,000	$993,000

In this example, the intrinsic bioremediation case appears significantly more expensive when calculated at a discount rate of 6 percent as might be used by a nonprofit entity. The conclusion would be that intrinsic bioremediation methods are expensive and not competitive and that it is more economical to perform a rapid, accelerated bioremediation, rather than a remediation that would extend over a longer period. Again, the developer might misread the market, and the user might overlook a potentially viable option if the second method of calculating the present cost were used.

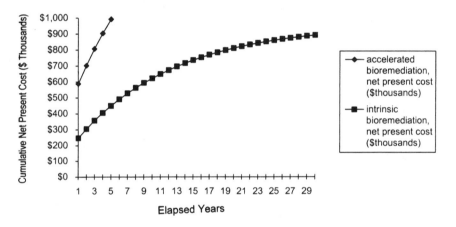

FIGURE 6-3   Net present cost comparison of hypothetical intrinsic versus accelerated bioremediation process. The calculations assume for accelerated bioremediation a $750,000 initial equipment cost and $200,000 per year for operation and maintenance over 5 years; for intrinsic bioremediation, the calculations assume a $300,000 initial equipment cost and operation and maintenance costs of $100,000 per year for 30 years.

The Environmental Protection Agency (EPA) should be responsible for establishing the data bases and appropriate formats for entering cost data. Then, every technology provider at a contaminated site where the federal government is involved (that is, every site governed by federal regulations such as Superfund and the Resource Conservation and Recovery Act and every federal facility undergoing cleanup) should be required to provide cost data for the data base as soon as the cleanup is under way. The EPA should advertise the data base and make it available electronically, on the Internet, as is already being done for technology assessments by the Ground Water Remediation Technologies Analysis Center.

FIELD IMPLEMENTATION COSTS

The guidelines presented in this chapter address primarily the ways in which first approximation costs can be developed to compare technologies, which can indicate whether a technology will be cost competitive or can be used to select candidate technologies for implementation. These techniques constitute the beginning steps in defining the final cost of implementing a technology in the field. As discussed in Chapter 3, the actual cost of implementation may differ from the pure technology costs for many reasons. Under ideal circumstances, these differences will have been addressed in preliminary evaluations, but often design changes are needed as more is learned about site conditions or as other unexpected costs arise. Estimation of costs for actual implementation is often an iterative process, which can result in costs different from those estimated in the selection process. In some cases, this can lead to a reevaluation of the initial technology selection, but in many cases the relative merits of the technology will still stand, particularly if all of the measures of success discussed in Chapter 4 were carefully considered.

CONCLUSIONS

Development of realistic cost data is essential for deciding whether a new technology will be cost competitive in the marketplace and for comparing candidate technologies at a particular site. Yet, figuring the costs of a technology can be a frustrating exercise. Currently, costs of technologies for cleaning up contaminated ground water and soil are not reported in formats that allow comparison of one technology to another or extrapolation of costs from one site to another. Inconsistent calculations and unstated assumptions made in estimating costs can remove a remediation technology from the menu of options being considered by a client. Problems with cost reporting include the following:

• Technology providers report costs using different metrics that cannot be compared. Costs reported using one type of measure, such as cost per area of containment wall installed, cannot be easily compared to another measure, such as cost per weight of contaminant contained.

• Technology providers often report only the up-and-running costs of using a technology, excluding variable costs such as permitting, mobilization of equipment, treatability studies, and system design. Failure to include variable costs may result in a technology appearing cost competitive when in fact it is not due to large variable costs.

• Methods used and assumptions made when computing costs are rarely reported. This lack of documentation makes it difficult for the technology user to judge the realism of cost data.

• There is no central data base where technology users can go to find consistent, comparable cost information about a range of innovative remediation technologies. Cost information developed by consultants is rarely compiled for general use.

RECOMMENDATIONS

To facilitate comparisons of the costs of different remediation technologies, the committee recommends several initiatives to standardize cost analysis and reporting and to improve the availability of remediation technology cost data:

•   **A working group composed of representative problem owners (corporations and government agencies) and technology developers should be convened under the auspices of an umbrella organization, such as the Remediation Technologies Development Forum or American Academy of Environmental Engineers, to develop and refine a standardized template system that can be used to compare the costs of different remediation technologies.** For contaminated ground water, a workable number of templates should be developed to represent the range of conditions of contaminant depth, aquifer thickness, and aquifer permeability. Similar templates should be developed for contaminated soil. Once the templates are developed and refined, federal agencies and private corporations should request that remediation technology vendors present cost data in the template format if the technology is to be evaluated for purchase. The templates can then be used to provide screening-level comparisons of remediation technologies designed to achieve the same level of public health and environmental protection. More detailed cost data, based on actual site conditions, would then need to be developed for the technologies that pass this first level of screening.

•   **Costs of remediation technologies should always be reported as cost per unit volume of the contaminated matrix treated, removed, or contained and as cost per mass of each specific contaminant removed, treated, or contained.** The starting concentration of the contaminant and the process rate should be provided along with the cost data. The amount of contaminated soil or ground water treated should also be reported because unit remediation costs can vary with the size of the operation.

•   **Cost estimates should include one-time start-up costs as well as the up-and-running cost of using the technology.** Start-up costs include the costs of site preparation, equipment mobilization, pilot testing, permitting, and system design.

•   **Table 6-4 should be used as the minimum base of cost elements to be included in technology cost comparisons.** Where appropriate, technology developers and consultants should structure reporting according to level 3 of the federal work breakdown structure, taking into consideration the elements of levels 4 and 5 of the remediation work breakdown structure.

• **The Federal Remediation Technologies Roundtable should reevaluate the role of the work breakdown structure in standardizing remediation cost reporting and should document the system in a way that facilitates understanding by the private sector.** The work breakdown structure may be too rigid in format to be appropriate for standardizing costs for the wide range of technologies to be encountered and may not be an efficient tool for the private sector to use in developing cost data for new technologies. The role of the work breakdown structure should be reevaluated and a guidance manual prepared to help the private sector use this tool. The instruction manual should be advertised to providers and users and should be available in an on-line version.

• **Assumptions about discount rates and tax benefits should be clearly stated in estimates of present costs of a technology that operates over an extended time period.** In developing cost estimates for technology users, technology providers should tailor their assumptions about discount rates and taxes to the needs of the user.

• **The EPA should extend its technology assessment initiatives to include a national data base for reporting the cost of remediation technologies.** For each technology, costs should be included for the template sites for which the technology would be appropriate. The data base should also list actual costs from sites where the technology is already in use according to weight of contaminant and volume of contaminated matrix removed, treated, or contained. Capital and operating costs should be reported separately, so that users can develop their own present cost estimates using discount rates appropriate to their own needs.

REFERENCES

Ellis, D. E. 1996. Intrinsic remediation in the industrial marketplace. In Proceedings of the Symposium on Natural Attenuation of Chlorinated Organics in Groundwater. EPA/154-R-96/509. Washington, D.C.: EPA.

Federal Remediation Technologies Roundtable. 1995. Guide to Documenting Cost and Performance for Remediation Projects. EPA-542-B-95-002. Washington, D.C.: EPA.

Herriksen, A. D., and S. R. Booth. 1995. Evaluating the cost-effectiveness of new environmental technologies. Remediation 5(1):7-24.

Stermole, F. J., and J. M. Stermole. 1996. Economic Evaluation and Investment Decision Methods, 9th Ed. Golden, Colo.: Investment Evaluations Corporation.

A
Data Bases Containing Information About Remediation Technologies

Data Base	Description/Information Provided	Organization/Contact
Alternative Treatment Technology Information Center (ATTIC) Network	Consists of the following online data bases:	Maintained by the Technical Support Branch of the EPA's Risk Reduction Engineering Laboratory (RREL). For information, contact the system operator at (513) 569-7272, or contact the system by Internet using Telnet or FTP at CINBDF.CIN.EPA.GOV.
• ATTIC Data Base	Contains abstracts and bibliographic citations to technical reports, bulletins, and other publications from EPA, other federal and state agencies, and industry focused on technologies for hazardous waste treatment with performance and cost data, quality assurance information, and contact information.	
• Risk Reduction Engineering Laboratory Treatability Data Base	Provides contaminant information, including physicochemical properties, environmental data, and applicability of treatment technologies.	
• Technology Assistance Directory	Lists experts from government, universities, and consulting firms providing guidance on technical issues or policy questions.	
• Calendar of Events	Lists national and international conferences, seminars, and workshops on treatment of hazardous waste.	
• Robert S. Kerr Environmental Laboratory Soil Transport and Fate Data Base	Provides data about the movement and fate of contaminants in soil matrices.	

continued on next page

• Hazardous Waste Collection Data Base	Includes reports, commercially published books, directives, and legislation on hazardous waste.	For information, contact Susan Richmond, Eastern Research Group, Inc., (617) 674-2851 (fax).
Bioremediation in the Field Search System (BFSS)	Provides information on ex situ and in situ bioremediation technologies at more than 160 sites nationwide. The data base includes full-scale remediation efforts and treatability and feasibility studies.	
Case Study Data System	Contains more than 200 case studies of RCRA sites, addresses, and topics such as floodplains, disposal technology, treatment, and environmental effects.	For information, contact Corrective Action Programs Branch, Office of Solid Waste, (703) 308-8484.
Clean-Up Information Bulletin Board System (CLU-IN)	Provides information about hazardous waste cleanup technologies and activities through online messaging, bulletin board service, and web site.	Maintained by the EPA's Technology Innovation Office. For information, contact the CLU-IN help line at (301) 589-8368 or visit their web site (http://clu-in.com).
Comprehensive Environmental Response, Compensation, and Liability Information System (CERCLIS)	Provides access to information on initial identification of potentially hazardous waste sites for listing on National Priorities List, including site location, classification, assessment data, remedial information, and points of contact.	Maintained on EPA's central computing system, the National Computer Center in Research Triangle Park, North Carolina. For information, contact Michael Cullen, CERCLIS help line, at (703) 603-8881.

Data Base	Description/Information Provided	Organization/Contact
Cost of Remedial Action Model (CORA)	This computerized expert model is designed to recommend remedial actions for Superfund hazardous waste sites and estimate the cost of these actions. The model is comprised of two independent systems: an expert system that uses site information to recommend a range of remedial response actions and a cost system that develops cost estimates for the technologies selected.	For information, contact the CORA hotline, CH$_2$M Hill, (703) 478-3566.
Defense Environmental Network and Information Exchange (DENIX)	Serves as a centralized communications platform for disseminating information pertaining to DOD's scheduled meetings, training, cleanup sites, and technologies.	For information, contact Kim Grein, Army Corps of Engineers, Civil Engineering Research Laboratory, (217) 352-6511 ext. 6790.
EPA Online Library System (OLS)	Contains bibliographic citations to EPA reports received from National Technical Information Service and regional EPA libraries, as well as books and articles. Records can be searched by title, authors, corporate sources, keywords, year of publication, and report number.	Maintained by the EPA's National Computer Center. For information, contact the EPA library at (202) 260-5922.
Environmental Technologies Remedial Actions Data Exchange (EnviroTRADE)	Contains data on international environmental restoration and waste management technologies, organizations, sites, activities, funding, and contracts.	Contact International Technology Exchange Program, DOE, (301) 903-7930, or visit the web site (http://www.gnet.org).
Environmental Technology Information System (TIS)	Offers advice on screening remedial options based on site-specific input information.	For information, contact Bill Lawrence, Idaho National Engineering Laboratory, (208) 526-1364.

Global Network for Environmental Technology (GNET)	Provides services to enhance efforts to communicate and exchange information to commercialize innovative environmental technologies. Offers Environmental Technology News Briefs, the Environment and Technology Business Forum, and Environmental and Technology Information.	For information, contact GNET Client Services by phone at (703)750-6401, or visit the web site at (http:// gnetnet.org).
Ground-Water Remediation Technologies Analysis Center (GWRTAC) Data Bases	Established to improve understanding and development of innovative ground water remediation technologies, support customer groups requiring access to technical information, and provide information transfer between developers and users. Searchable online databases on laboratory/bench-scale projects, controlled field demonstrations, and commercial application are available for scanning, uploading, and downloading.	For information, contact GWRTAC by telephone at (800) 373-1973 or by email at gwrtac@chmr.com. The web site address is http://www.gwrtac.org.
Hazardous Waste Superfund Collection Data Base	Includes bibliographic references and abstracts on EPA reports, policy and guidance directives, legislation, regulations, and nongovernment books.	For information, contact Hazardous Waste Superfund Collection, EPA Headquarters Library, at (202) 260-5934, or visit the CLU-IN website (http://clu-in.com).
Records of Decision Data Base	Contains the full text of all signed and published Superfund records of decision and can be searched by indexed fields, including site location, contaminated media, key contaminants, and selected remedy.	Maintained at EPA's National Computer Center. For information, contact the CERCLIS help line at (703) 603-8881.

continued on next page

Data Base	Description/Information Provided	Organization/Contact
Remediation Information Management System (RIMS)	On-line library of information about more than 800 existing and emerging remediation technologies; includes case studies, third-party evaluations, peer-reviewed articles, and technical and cost reports.	For information, contact Charles Corbin, Remedial Technologies Network, L.L.C., (703) 481-9095, or visit the web site (http://www.remedial.com).
ReOpt Data Bases	Developed by the Department of Energy as a part of the Remedial Action Assessment System, ReOpt contains information about potential technologies for cleanup at DOE or other waste sites, auxiliary information about possible hazardous or radioactive contaminants, and applicable federal regulations governing disposal.	Maintained by the DOE's Pacific Northwest Laboratory. For information, contact Janet Bryant at (509) 375-3765.
Resource Conservation and Recovery Information System (RCRIS)	Contains information collected by EPA regional offices and states from permit applications, notification forms, and inspection reports on facilities that handle hazardous waste and supports permit writing and enforcement activities of the EPA corrective action program.	Maintained at the EPA's National Computer Center. For information, contact (703) 308-7914 or visit the web site (http://www.rtk.net).
Soil Transport and Fate Data Base and Model Management System	The data base provides information on chemical properties, toxicity, transformation, and bioaccumulation for chemical compounds; includes information on approximately 400 chemicals and models for predicting fate and transport in the vadose zone.	For information, contact David S. Burden, R. S. Kerr Environmental Research Laboratory at (405) 436-8606, or call (405) 436-8586 for technical support.

TechDirect	Highlights new publications and events of interest to site remediation and site assessment professionals.	For information, call the EPA's Technology Innovation Office at (703) 603-9910, or send an e-mail to TechDirect.TI0@epamail.epa.gov.
Technology Access Services	Provides information on newest technologies available through the research and development programs of the federal government laboratories	For information, contact Gateway/Technology Access Services at (800)-678-6882 or (304) 243-2551.
Vendor Information System for Innovative Treatment Technologies (VISITT)	Contains information provided by vendors of innovative remediation technologies. Major technology categories include acid/solvent extraction, bioremediation, chemical treatment, soil vapor extraction, soil washing, thermal desorption, and vitrification.	For information, contact the VISITT help line at (800) 245-4505 or (703) 883-8448, or see the CLU-IN web site (http://clu-in.com).

SOURCE: Adapted from Federal Remediation Technologies Roundtable, 1995, Accessing Federal Data Bases for Contaminated Site Clean-Up Technologies, Fourth Edition, Washington, D.C.: Environmental Protection Agency.

Biographical Sketches of Committee Members and Staff

COMMITTEE MEMBERS

P. Suresh Rao, *chair*, is a graduate research professor in the Soil and Water Science Department of the University of Florida. His research interests include the development and field testing of process-level models for predicting the fate of pollutants in soils and ground water. He also has worked extensively on the field testing of innovative in situ flushing technologies for site remediation. He has worked with state and federal agencies in providing scientific bases for environmental regulatory policy. Dr. Rao is a former WSTB member and member of the WSTB's Committee on Ground Water Vulnerability Assessment and Committee on Ground Water Modeling Assessment. He received a Ph.D. in soil physics from the University of Hawaii.

Richard Brown, *vice-chair*, is vice-president of remediation technology for Fluor Daniel GTI. His responsibilities include the development and implementation of remediation technologies such as bioremediation, soil vapor extraction, and air sparging. Before joining Fluor Daniel GTI, Dr. Brown was director of business development for Cambridge Analytical Associates' Bioremediation Systems Division and technology manager for FMC Corporation's Aquifer Remediation Systems. Dr. Brown holds patents on applications of bioreclamation technology, on the use of hydrogen peroxide in bioreclamation, and on an improved nutrient formulation for the biological treatment of hazardous wastes. He was a member of the WSTB's Committee on In Situ Bioremediation. Dr. Brown received a B.A. in chemistry from Harvard and a Ph.D. in inorganic chemistry from Cornell University.

Richelle Allen-King is an assistant professor in the Department of Geology at Washington State University, where she teaches courses in ground water and contaminant fate and transport. Her research focuses on the biogeochemistry of contaminants in the subsurface environment. She is currently a member of the science advisory board for Washington State's environmental regulatory agency. She received a B.A. in chemistry from the University of California, San Diego, and a Ph.D. in earth sciences (hydrogeology) from the University of Waterloo.

William Cooper is chair of the Department of Chemistry at the University of North Carolina at Wilmington. Prevously, he was director of the Drinking Water Research Center at Florida International University. His current research focuses on high-energy electron beam irradiation of contaminated water, a process tested under the Superfund Innovative Technology Evaluation Program. Dr. Cooper is an environmental chemist, with a Ph.D. in marine and atmospheric chemistry from the University of Miami.

Wilford Gardner, a member of the National Academy of Sciences, recently re-tired as dean of the College of Natural Resources at the University of California, Berkeley. His research has focused on movement of fluids in porous media, soil physics, soil moisture measurement, and environmental physics. He has been a National Science Foundation senior fellow at Cambridge University and a Fulbright Lecturer at the University of Ghent. He was a member of the WSTB's Committee on Irrigation-Induced Water Quality Problems and is currently a mem-ber of the WSTB. He received a Ph.D. in physics from Iowa State College.

Michael Gollin is a partner at Spencer & Frank in Washington, D.C. He is a registered patent attorney with experience in environmental law and litigation. He has built an international practice helping clients to protect, enforce, defend, and market intellectual property, with an emphasis on biotechnology and envi-ronmental technology. He holds an A.B. in biochemical sciences from Princeton, an M.S. in zoology and molecular biology from the University of Zurich, and a J.D. from Boston University.

Thomas Hellman is vice-president for environmental affairs at Bristol-Myers Squibb Company. During his career, he has managed environmental health and safety operations for a range of companies, including General Electric, Allied Chemical, and Union Carbide. He served on the WSTB's Committee on Ground Water Quality Protection and is currently a member of the WSTB. He received a B.A. in chemistry from Williams College and a Ph.D. in organic chemistry from Pennsylvania State University.

Diane Heminway is the western New York coordinator for the Citizens' Envi-ronmental Coalition, a statewide coalition of 90 environmental, community, and labor organizations in New York State. Heminway became active in ground water cleanup issues as a result of a pesticide spill near her children's elementary school. She is very familiar with public concerns about the limitations of cleanup technologies and has been outspoken about the need for governments and companies to present citizens with complete information about the capabilities of technologies. She currently serves on the Water Management Advisory Committee and the Working Group of the Pesticide Management Advisory Board of the New York State Department of Environmental Conservation.

Richard Luthy is a professor in (and former head of) the Department of Civil and Environmental Engineering at Carnegie Mellon University. In addition to academic responsibilities, he has consulted on a range of waste treatment and remediation issues for both the public and private sectors. His research interests in environmental engineering include physicochemical processes for industrial waste reduction and treatment, remediation of contaminated soil using physicochemical and microbial processes, and applied aquatic chemistry. Dr. Luthy is a former president of the Association of Environmental Engineering Professors, and a past chair of the Gordon Research Conference on Environmental Sciences. He received a B.S. in chemical engineering and an M.S. and Ph.D. in environmental engineering from the University of California, Berkeley.

Roger Olsen is vice-president and senior geochemist for Camp Dresser & McKee. He is responsible for project management and technical supervision of geochemical and hazardous waste investigations. His experience includes design of sampling and analytical programs; evaluation of risks and impacts; evaluation of treatment and disposal options; implementation of quality control procedures; and design and engineering of hazardous waste disposal and remediation programs. He has expertise in the mobility, degradation, and transport of metals and organic compounds in soil water systems. He received a B.S. in mineral engineering chemistry and a Ph.D. in geochemistry from the Colorado School of Mines.

Philip Palmer, a senior environmental fellow in the DuPont Chemicals Core Resources Section of the Corporate Remediation Group, has over 15 years of experience in the field of remediation technology development. He currently heads a group of 40 that is evaluating remediation technologies. Palmer oversees development and pilot testing of new technologies on DuPont sites and assessment of the company's remediation technology needs. Mr. Palmer served as a leader and member of the Chemical Manufacturers Association RCRA Regulations Task Force from the inception of RCRA until 1990. He is a former member of the National Research Council's Commission on Geosciences, Environment, and Resources. He holds a B.S. and an M.S. in chemical engineering from Cornell University. He holds an M.S. in environmental engineering from Drexel University.

Frederick Pohland, a member of the National Academy of Engineering, is Edward R. Weidlein Chair of Environmental Engineering and professor of civil engineering at the University of Pittsburgh. His research interests include water and waste chemistry and microbiology, solid and hazardous waste management, and environmental impact monitoring and assessment. He earned an M.S. in civil engineering and a Ph.D. in environmental engineering from Purdue University.

Ann Rappaport is currently an assistant professor in the Department of Civil and Environmental Engineering at Tufts University and director of the university's Hazardous Materials Management Program. Previously, she served as chief of policy and program development for the Division of Hazardous Waste, Massachusetts Department of Environmental Quality Engineering. She earned a B.A. in Asian and environmental studies from Wellesley, an M.S. in civil engineering from Massachusetts Institute of Technology, and a Ph.D. in civil engineering from Tufts University.

Martin Sara is principal hydrogeologist for RUST Environment & Infrastructure. His current responsibilities include conducting assessments of hazardous waste sites and managing monitoring programs for solid and hazardous waste landfills. He is active with the American Society for Testing of Materials (ASTM). He authored ASTM Standard D5092, "Design and Installation of Ground-water Monitoring Wells in Aquifers." He currently chairs two ASTM committees: the Monitoring Wells Design and Construction committee and the Geochronology and Environmental Isotopes committee. Mr. Sara recently authored a text, "Standard Handbook for Solid and Hazardous Waste Facility Assessments," published by Lewis Publishers and used in the EPA's Superfund University Training Institute. He holds a B.S. in geology from the University of Illinois and an M.S. in geological sciences from the University of Southern California.

Dag Syrrist is a partner with the venture capital firm Vision Capital in Boston, Massachusetts. Previously, he was manager of environmental operations and the principal industry liaison for the Environmental Finance Group at Technology Funding. His responsibilities included establishing the industry and government relationships necessary to implement Technology Funding's environmental investment strategies, including technology transfer, corporate alliances, and licensing. Mr. Syrrist also acted as Technology Funding's primary coordinator with the Environmental Protection Agency, Department of Defense, and national laboratories. He has served on several federal, state, and regional advisory committees focusing on technology development, diffusion, and financing. Mr. Syrrist holds a B.A. in business administration from Lincoln University and an M.A. in international economics from San Francisco State University.

Brian Wagner is a research hydrologist in the U.S. Geological Survey's National Research Program. His research interests include data network design for environmental monitoring and assessment, experimental design for understanding contaminant fate and transport, and optimization techniques for water resources management. He received a B.S. in civil engineering from Drexel University and an M.S. and a Ph.D. in applied hydrogeology from Stanford University.

STAFF

Jacqueline A. MacDonald is associate director of the National Research Council's Water Science and Technology Board. She directed the studies that led to the reports *Alternatives for Ground Water Cleanup, In Situ Bioremediation: When Does It Work?, Safe Water From Every Tap: Improving Water Service to Small Communities*, and *Freshwater Ecosystems: Revitalizing Educational Programs in Limnology*. She received the 1996 National Research Council Award for Distinguished Service. Ms. MacDonald earned an M.S. degree in environmental science in civil engineering from the University of Illinois, where she received a university graduate fellowship and Avery Brundage scholarship, and a B.A. degree magna cum laude in mathematics from Bryn Mawr College.

Angela F. Brubaker is a research assistant at the National Research Council's Water Science and Technology Board. She prepared the report of the Committee on Innovative Remediation Technologies for publication and assisted with editing the final draft. She received a B.A. in liberal arts from Eastern Mennonite College in 1990.

Ellen A. de Guzman is a project assistant at the National Research Council's Water Science and Technology Board. She assisted in preparing the final draft of the report. She received a B.A. from the University of the Philippines.

Index

S